# Holt Mathematics

## Chapter 5 Resource Book

**HOLT, RINEHART AND WINSTON**

A Harcourt Education Company

Orlando • Austin • New York • San Diego • London

Copyright © by Holt, Rinehart and Winston

All rights reserved. No part of this publication may be reproduced or transmitted in any form or by any means, electronic or mechanical, including photocopy, recording, or any information storage and retrieval system, without permission in writing from the publisher.

Teachers using HOLT MATHEMATICS may photocopy complete pages in sufficient quantities for classroom use only and not for resale.

Printed in the United States of America

If you have received these materials as examination copies free of charge, Holt, Rinehart and Winston retains title to the materials and they may not be resold. Resale of examination copies is strictly prohibited and is illegal.

Possession of this publication in print format does not entitle users to convert this publication, or any portion of it, into electronic format.

ISBN 0-03-078221-X

7 8 9    170    10 09

# CONTENTS

**Blackline Masters**

| | |
|---|---|
| Parent Letter | 1 |
| Lesson 5-1 Practice A, B, C | 3 |
| Lesson 5-1 Reteach | 6 |
| Lesson 5-1 Challenge | 7 |
| Lesson 5-1 Problem Solving | 8 |
| Lesson 5-1 Reading Stratagies | 9 |
| Lesson 5-1 Puzzles, Twisters & Teasers | 10 |
| Lesson 5-2 Practice A, B, C | 11 |
| Lesson 5-2 Reteach | 14 |
| Lesson 5-2 Challenge | 15 |
| Lesson 5-2 Problem Solving | 16 |
| Lesson 5-2 Reading Stratagies | 17 |
| Lesson 5-2 Puzzles, Twisters & Teasers | 18 |
| Lesson 5-3 Practice A, B, C | 19 |
| Lesson 5-3 Reteach | 22 |
| Lesson 5-3 Challenge | 23 |
| Lesson 5-3 Problem Solving | 24 |
| Lesson 5-3 Reading Stratagies | 25 |
| Lesson 5-3 Puzzles, Twisters & Teasers | 26 |
| Lesson 5-4 Practice A, B, C | 27 |
| Lesson 5-4 Reteach | 30 |
| Lesson 5-4 Challenge | 31 |
| Lesson 5-4 Problem Solving | 32 |
| Lesson 5-4 Reading Stratagies | 33 |
| Lesson 5-4 Puzzles, Twisters, & Teasers | 34 |
| Lesson 5-5 Practice A, B, C | 35 |
| Lesson 5-5 Reteach | 38 |
| Lesson 5-5 Challenge | 39 |
| Lesson 5-5 Problem Solving | 40 |
| Lesson 5-5 Reading Stratagies | 41 |
| Lesson 5-5 Puzzles, Twisters & Teasers | 42 |
| Lesson 5-6 Practice A, B, C | 43 |
| Lesson 5-6 Reteach | 46 |
| Lesson 5-6 Challenge | 47 |
| Lesson 5-6 Problem Solving | 48 |
| Lesson 5-6 Reading Stratagies | 49 |
| Lesson 5-6 Puzzles, Twisters & Teasers | 50 |
| Lesson 5-7 Practice A, B, C | 51 |
| Lesson 5-7 Reteach | 54 |
| Lesson 5-7 Challenge | 55 |
| Lesson 5-7 Problem Solving | 56 |
| Lesson 5-7 Reading Stratagies | 57 |
| Lesson 5-7 Puzzles, Twisters & Teasers | 58 |
| Lesson 5-8 Practice A, B, C | 59 |
| Lesson 5-8 Reteach | 62 |
| Lesson 5-8 Challenge | 63 |
| Lesson 5-8 Problem Solving | 64 |
| Lesson 5-8 Reading Stratagies | 65 |
| Lesson 5-8 Puzzles, Twisters & Teasers | 66 |
| Lesson 5-9 Practice A, B, C | 67 |
| Lesson 5-9 Reteach | 70 |
| Lesson 5-9 Challenge | 71 |
| Lesson 5-9 Problem Solving | 72 |
| Lesson 5-9 Reading Stratagies | 73 |
| Lesson 5-9 Puzzles, Twisters & Teasers | 74 |
| Lesson 5-10 Practice A, B, C | 75 |
| Lesson 5-10 Reteach | 78 |
| Lesson 5-10 Challenge | 79 |
| Lesson 5-10 Problem Solving | 80 |
| Lesson 5-10 Reading Stratagies | 81 |
| Lesson 5-10 Puzzles, Twisters & Teasers | 82 |
| Answers to Blackline Masters | 83 |

Copyright © by Holt, Rinehart and Winston.
All rights reserved.

Holt Mathematics

Date _____

Dear Family,

In this chapter, your child will learn how to compute with fractions. These fraction skills are essential for problem solving and for future success in algebra.

Fractions are "unlike" when they have different denominators, such as $\frac{1}{3}$ and $\frac{1}{5}$. To add or subtract unlike fractions, you must first rewrite them as equivalent fractions with a **common denominator.** The least common denominator is the smallest number that is a multiple of both numbers. It is also called the **least common multiple** (LCM).

To find the least common multiple of 3 and 5, you can make a list of the multiples of each number.

Multiples of 3 are 3, 6, 9, 12, 15, 18, 21, 24, 27, 30.

Multiples of 5 are 5, 10, 15, 20, 25, 30, 35, 40, 45.

The common multiples in the lists are 15 and 30. The smallest is 15, so 15 is the **least common denominator.**

To add $\frac{1}{3}$ and $\frac{1}{5}$:

$\frac{1}{3} + \frac{1}{5}$     Find the least common denominator for 3 and 5. 15 is the least common denominator.

$\frac{1}{3} = \frac{5}{15}$
$+ \frac{1}{5} = \frac{3}{15}$     Write each fraction with the common denominator.

$\frac{5}{15} + \frac{3}{15}$     Add the numerators. Keep the denominator.
$= \frac{8}{15}$

To multiply with fractions, your child will do the following:

$\frac{1}{3} \times \frac{3}{5}$     Multiply numerators. Multiply denominators.

$= \frac{3}{15}$     The greatest common factor of 3 and 15 is 3, so each is divided by 3 to simplify the fraction.

$= \frac{1}{5}$     This is the answer in simplest form.

**Holt Mathematics**

In order to multiply with a mixed number, students will first write the mixed number as an improper fraction.

$\frac{1}{3} \times 1\frac{1}{2}$

$\frac{1}{3} \times \frac{3}{2}$     Write $1\frac{1}{2}$ as $\frac{3}{2}$.

$\frac{1}{3} \times \frac{3}{2}$     Multiply numerators. Multiply denominators.

$= \frac{3}{6}$     The greatest common factor of 3 and 6 is 3, so each is divided by 3 to simplify the fraction.

$= \frac{1}{2}$     This is the answer in simplest form.

To divide fractions, **reciprocals** are used. The fractions $\frac{2}{3}$ and $\frac{3}{2}$ are reciprocals because the numerators and denominators are reversed. The reciprocal of a whole number, such as 8, is one over the number, or $\frac{1}{8}$.

To divide with fractions, you can multiply by the reciprocal of the divisor.

$\frac{4}{5} \div 5$

$= \frac{4}{5} \times \frac{1}{5}$     Rewrite as multiplication.

     Use the reciprocal of 5, $\frac{1}{5}$.

$= \frac{4}{25}$     Multiply.

Reciprocals are also used to solve fraction equations.

$\frac{2}{3}n = 14$

$\frac{2}{3}n \div \frac{2}{3} = 14 \div \frac{2}{3}$     Divide both sides of the equation by $\frac{2}{3}$.

$\frac{2}{3}n \times \frac{3}{2} = 14 \times \frac{3}{2} = \frac{14}{1} \times \frac{3}{2}$     Multiply by $\frac{3}{2}$ the reciprocal of $\frac{2}{3}$.

$n = \frac{42}{2}$

$n = 21$

For additional resources, visit go.hrw.com and enter the keyword MR7 Parent.

Name _____ Date _____ Class _____

## Practice A
**LESSON 5-1** *Least Common Multiple*

**List the first five multiples.**

**1.** 2

**2.** 6

**3.** 12

**4.** 3

**5.** 7

**6.** 10

**Find the least common multiple (LCM).**

**7.** 2 and 3

2: _____

3: _____

**8.** 2 and 8

2: _____

8: _____

**9.** 2 and 4

2: _____

4: _____

**10.** 2, 3, and 4

2: _____

3: _____

4: _____

**11.** 3, 4, and 6

3: _____

4: _____

6: _____

**12.** 3, 5, and 10

3: _____

5: _____

10: _____

**13.** 2, 4, and 5

**14.** 2, 4, and 6

**15.** 2, 3, and 6

**16.** Hot dogs come in packs of 8. Hot dog rolls come in packs of 12. What is the least number of packs of each Shawn should buy to have enough to serve 24 people and have none left over?

**17.** Debbie wants to invite 60 people to her party. Invitations come in packs of 12 and stamps come in sheets of 10. What is the least number of each she should buy to mail an invitation to each person and have no supplies left over?

Holt Mathematics

Name _____ Date _____ Class _____

## LESSON 5-1 Practice B
### Least Common Multiple

Find the least common multiple (LCM).

1. 2 and 5

2. 4 and 3

3. 6 and 4

4. 6 and 8

5. 5 and 9

6. 4 and 5

7. 10 and 15

8. 8 and 12

9. 6 and 10

10. 3, 6, and 9

11. 2, 5, and 10

12. 4, 7, and 14

13. 3, 5, and 9

14. 2, 5, and 8

15. 3, 9, and 12

16. Mr. Stevenson is ordering shirts and hats for his Boy Scout troop. There are 45 scouts in the troop. Hats come in packs of 3, and shirts come in packs of 5. What is the least number of packs of each he should order to so that each scout will have 1 hat and 1 shirt, and none will be left over?

17. Tony wants to make 36 party bags. Glitter pens come in packs of 6. Stickers come in sheets of 4, and balls come in packs of 3. What is the least number of each package he should buy to have 1 of each item in every party bag, and no supplies left over?

18. Glenda is making 30 school supply baskets. Notepads come in packs of 5. Erasers come in packs of 15, and markers come in packs of 3. What is the least number of each package she should buy to have 1 of each item in every basket, and no supplies left over?

Name _____ Date _____ Class _____

## LESSON 5-1 Practice C
### Least Common Multiple

**Find the least common multiple (LCM).**

1. 6 and 9

2. 6 and 10

3. 12 and 8

_____ _____ _____

4. 5 and 13

5. 9 and 12

6. 11 and 12

_____ _____ _____

7. 4, 7, and 14

8. 5, 12, and 15

9. 8, 14, and 16

_____ _____ _____

10. 6, 8, and 16

11. 4, 8, and 64

12. 6, 10, and 12

_____ _____ _____

13. 3, 6, 9, and 12

14. 4, 6, 8, and 10

15. 2, 6, 8, and 12

_____ _____ _____

16. Mr. Simon wants to make packages of art supplies for his students. Pads of paper come 4 to a box, pencils come 27 to a box, and erasers come 12 to a box. What is the least number of kits he can make if he wants each kit to be the same and he wants no supplies left over? How many boxes of paper must he buy? how many boxes of pencils? how many boxes of erasers?

_____

17. Find the LCM and the GCF of 48 and 72. Now find the product of 48 and 72 and the product of the GCF and LCM. Describe the relationship between the two products. This relationship is true for all whole numbers. How could you use this relationship to solve problems?

_____

_____

_____

Name _____ Date _____ Class _____

## LESSON 5-1 Reteach
### Least Common Multiple

The smallest number that is a multiple of two or more numbers is called the least common multiple (LCM).

To find the least common multiple of 3, 6, and 8, list the multiples for each number and put a circle around the LCM in the three lists.

Multiples of 3: 3, 6, 9, 12, 15, 18, 21, (24)
Multiples of 6: 6, 12, 18, (24), 30, 36, 42
Multiples of 8: 8, 16, (24), 32, 40, 48, 56

So 24 is the LCM of 3, 6, and 8.

**List the multiples of each number to help you find the least common multiple of each group.**

**1.** 3 and 4

Multiples of 3: _____

Multiples of 4: _____

LCM: _____

**2.** 5 and 7

Multiples of 5: _____

Multiples of 7: _____

LCM: _____

**3.** 8 and 12

Multiples of 8: _____

Multiples of 12: _____

LCM: _____

**4.** 2 and 9

Multiples of 2: _____

Multiples of 9: _____

LCM: _____

**5.** 4 and 6

Multiples of 4: _____

Multiples of 6: _____

LCM: _____

**6.** 4 and 10

Multiples of 4: _____

Multiples of 10: _____

LCM: _____

**7.** 2, 5, and 6

Multiples of 2: _____

Multiples of 5: _____

Multiples of 6: _____

LCM: _____

**8.** 3, 4, and 9

Multiples of 3: _____

Multiples of 4: _____

Multiples of 9: _____

LCM: _____

**9.** 8, 10, and 12

Multiples of 8: _____

Multiples of 10: _____

Multiples of 12: _____

LCM: _____

Copyright © by Holt, Rinehart and Winston.
All rights reserved.

Holt Mathematics

Name _____ Date _____ Class _____

## LESSON 5-1 Challenge
### Moons Over Neptune

We measure one month by our moon's orbital period, or the time it takes the Moon to travel once around Earth, which is about 30 days. But what if you lived on Neptune? It has 8 moons! How could you pick just one moon to measure your months? One possible solution is to calculate one month based on when two of Neptune's moons are in conjunction at some arbitrary starting point in the sky, or appear to be in the same place in the sky. The diagram below shows some of the moons you could use to measure your months on Neptune.

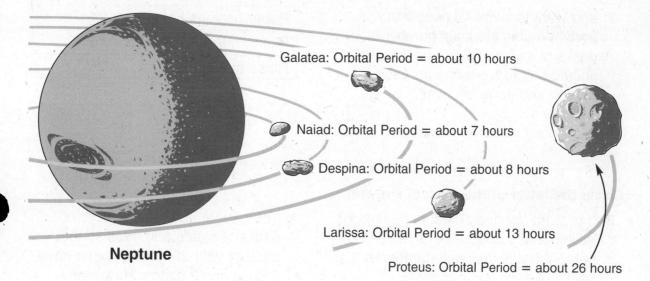

Galatea: Orbital Period = about 10 hours
Naiad: Orbital Period = about 7 hours
Despina: Orbital Period = about 8 hours
Larissa: Orbital Period = about 13 hours
Proteus: Orbital Period = about 26 hours

**Neptune**

**Use the diagram and least common multiples to complete the chart below. For each row, write how long your month on Neptune would be if you used those moons in conjunction as the length of one month.**

| Neptune Moons to Use | Length of One Neptune Month |
|---|---|
| Naiad and Despina | |
| Larissa and Proteus | |
| Galatea and Despina | |
| Despina and Proteus | |

## Problem Solving
### 5-1 Least Common Multiple

**Use the table to answer the questions.**

1. You want to have an equal number of plastic cups and paper plates. What is the least number of packs of each you can buy?

   _____

   _____

2. You want to invite 48 people to a party. What is the least number of packs of invitations and napkins you should buy to have one for each person and none left over?

   _____

   _____

**Party Supplies**

| Item | Number per Pack |
|---|---|
| Invitations | 12 |
| Balloons | 30 |
| Paper plates | 10 |
| Paper napkins | 24 |
| Plastic cups | 15 |
| Noise makers | 5 |

**Circle the letter of the correct answer.**

3. You want to have an equal number of noisemakers and balloons at your party. What is the least number of packs of each you can buy?

   A  1 pack of balloons and 1 pack of noise makers
   B  1 pack of balloons and 2 packs of noise makers
   C  1 pack of balloons and 6 packs of noise makers
   D  6 packs of balloons and 1 pack of noise makers

4. You bought an equal number of packs of plates and cups so that each of your 20 guests would have 3 cups and 2 plates. How many packs of each item did you buy?

   F  1 pack of cups and 1 pack of plates
   G  3 packs of cups and 4 packs of plates
   H  4 packs of cups and 3 packs of plates
   J  4 packs of cups and 4 packs of plates

5. The LCM for three items listed in the table is 60 packs. Which of the following are those three items?

   A  balloons, plates, noise makers
   B  noise makers, invitations, balloons
   C  napkins, cups, plates
   D  balloons, napkins, plates

6. To have one of each item for 120 party guests, you buy 10 packs of one item and 24 packs of the other. What are those two items?

   F  plates and invitations
   G  balloons and cups
   H  napkins and plates
   J  invitations and noise makers

Name _____ Date _____ Class _____

## LESSON 5-1 Reading Strategies
### Understanding Vocabulary

**Least** means the smallest in size. The person with the least amount of homework has the smallest amount of work to do.

**Common** means shared. You may have classes in common with some of your friends.

A **multiple** is the answer to a multiplication problem.

The multiples of 5 are the answers to multiplying numbers by 5.

$1 \times 5 = 5 \quad 2 \times 5 = 10 \quad 3 \times 5 = 15 \quad 4 \times 5 = 20$

The **least common multiple** is the smallest multiple two numbers have in common.

**Follow the steps for finding the least common multiple of 5 and 10.**

1. List the first 10 multiplies of 5.

   _____

2. List the first 5 multiples of 10.

   _____

3. What multiples do 5 and 10 have in common?

   _____

4. Write the smallest multiple that 5 and 10 have in common. _____

5. What is the least common multiple of 5 and 10? _____

6. To find the least common multiple of two numbers, what is the first thing you should do?

   _____

7. What should you do next?

   _____

8. How do you know which of the common multiples is the least common multiple?

   _____

Name _____ Date _____ Class _____

## LESSON 5-1 Puzzles, Twisters & Teasers
### Math Abbreviation

Draw a line from each pair of numbers to common multiples for the numbers. Sometimes you will need to draw two lines from the same pair of numbers.

When you have finished, you will see a famous math abbreviation.

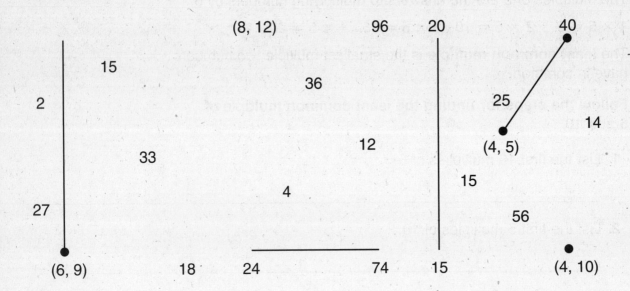

Copyright © by Holt, Rinehart and Winston.
All rights reserved.

Holt Mathematics

Name _____ Date _____ Class _____

## LESSON 5-2 Practice A
### Adding and Subtracting with Unlike Denominators

**Write the least common denominator for each pair of fractions.**

1. $\frac{1}{2}, \frac{2}{4}$

2. $\frac{1}{8}, \frac{2}{3}$

3. $\frac{1}{6}, \frac{1}{4}$

_____

4. $\frac{1}{3}, \frac{1}{5}$

5. $\frac{1}{5}, \frac{3}{4}$

6. $\frac{1}{5}, \frac{7}{10}$

_____

**Add or subtract. Write each answer in simplest form.**

7. $\frac{1}{2} + \frac{2}{3}$

8. $\frac{1}{2} - \frac{1}{4}$

9. $\frac{3}{4} - \frac{2}{3}$

_____

10. $\frac{2}{5} - \frac{1}{10}$

11. $\frac{1}{6} + \frac{1}{3}$

12. $\frac{1}{5} + \frac{7}{10}$

_____

13. $\frac{5}{8} - \frac{1}{4}$

14. $\frac{1}{5} + \frac{1}{4}$

15. $\frac{1}{2} - \frac{3}{8}$

_____

16. $\frac{2}{7} - \frac{1}{14}$

17. $\frac{3}{5} + \frac{1}{15}$

18. $\frac{5}{6} + \frac{1}{2}$

_____

19. Alice practices the piano $\frac{3}{4}$ hour every day. Today, however, she practiced for $\frac{1}{2}$ hour longer than usual. How long did Alice practice the piano today?

_____

20. One lap around the school's track is $\frac{1}{4}$ mile. Tyler ran two times around the track. Then he ran $\frac{5}{6}$ mile home. How far did Tyler run in all?

_____

Copyright © by Holt, Rinehart and Winston.
All rights reserved.

Holt Mathematics

Name _____ Date _____ Class _____

## LESSON 5-2 Practice B
### Adding and Subtracting with Unlike Denominators

Add or subtract. Write each answer in simplest form.

1. $\dfrac{6}{7} + \dfrac{1}{3}$

2. $\dfrac{3}{7} - \dfrac{2}{5}$

3. $\dfrac{1}{4} + \dfrac{3}{8}$

_____ _____ _____

4. $\dfrac{7}{8} - \dfrac{2}{3}$

5. $\dfrac{1}{6} + \dfrac{3}{5}$

6. $\dfrac{5}{6} - \dfrac{2}{3}$

_____ _____ _____

7. $\dfrac{5}{9} - \dfrac{1}{3}$

8. $\dfrac{7}{8} + \dfrac{3}{4}$

9. $\dfrac{5}{12} - \dfrac{1}{6}$

_____ _____ _____

10. $\dfrac{4}{5} - \dfrac{7}{11}$

11. $\dfrac{4}{9} + \dfrac{5}{6}$

12. $\dfrac{5}{8} + \dfrac{2}{3}$

_____ _____ _____

Evaluate each expression for $b = \dfrac{1}{3}$. Write your answer in simplest form.

13. $b + \dfrac{5}{8}$

14. $\dfrac{7}{9} - b$

15. $\dfrac{2}{7} + b$

_____ _____ _____

16. $b + b$

17. $\dfrac{11}{12} - b$

18. $\dfrac{3}{4} - b$

_____ _____ _____

19. There are three grades in Kyle's middle school—sixth, seventh, and eighth. One-third of the students are in sixth grade and $\dfrac{1}{4}$ are in seventh grade. What fraction of the schools' students are in eighth grade?

_____

20. Sarah is making a dessert that calls for $\dfrac{4}{5}$ cup of crushed cookies. If she has already crushed $\dfrac{7}{10}$ cup, how much more does she need?

_____

Name _____ Date _____ Class _____

## LESSON 5-2 Practice C
### Adding and Subtracting with Unlike Denominators

**Evaluate. Write each answer in simplest form.**

1. $\dfrac{11}{12} + \dfrac{3}{5}$

2. $\dfrac{7}{12} - \dfrac{5}{16}$

3. $\dfrac{5}{6} + \dfrac{3}{10}$

4. $\dfrac{3}{4} - \dfrac{3}{14}$

5. $\dfrac{1}{2} + \dfrac{5}{17}$

6. $\dfrac{4}{5} - \dfrac{2}{9}$

7. $\dfrac{7}{8} - \dfrac{5}{12}$

8. $\dfrac{3}{16} + \dfrac{5}{6}$

9. $\dfrac{3}{16} + \dfrac{5}{32}$

10. $\dfrac{11}{12} - \dfrac{4}{9} + \dfrac{1}{2}$

11. $\dfrac{2}{15} + \dfrac{7}{25} - \dfrac{2}{5}$

12. $\dfrac{3}{14} - \dfrac{1}{8} + \dfrac{4}{7}$

**Evaluate each expression for $b = \dfrac{2}{5}$. Write your answer in simplest form.**

13. $b + \dfrac{9}{14}$

14. $\dfrac{7}{12} - b$

15. $\dfrac{11}{16} - b$

16. $b + \dfrac{8}{11}$

17. $\dfrac{4}{7} - b$

18. $\dfrac{14}{15} + b$

19. Ben, Shaneeka, and Phil live on the same street. Ben lives $\dfrac{6}{11}$ mile north of Phil, and $\dfrac{1}{3}$ mile north of Shaneeka. How far does Shaneeka live from Phil?

20. At the frog-jumping contest, Trevor's frog jumped $\dfrac{5}{6}$ foot and then $\dfrac{5}{9}$ foot. Mei's frog jumped $\dfrac{4}{5}$ foot and then $\dfrac{1}{2}$ foot. Whose frog jumped the farthest in all? How much farther?

# Reteach

## 5-2 Adding and Subtracting with Unlike Denominators

Unlike fractions have different denominators. To add and subtract fractions, you must have a common denominator. The least common denominator (LCD) is the least common multiple of the denominators.

To add or subtract unlike fractions, first find the LCD of the fractions.

$\frac{2}{3} + \frac{1}{4}$

Multiples of 4: 4, 8, **12**,...
Multiples of 3: 3, 6, 9, **12**,...
The LCD is 12.

Next, use fraction strips to find equivalent fractions.

Then use fraction strips to find the sum or difference.

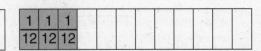

$\frac{8}{12} + \frac{3}{12} = \frac{11}{12}$

So, $\frac{2}{3} + \frac{1}{4} = \frac{11}{12}$.

**Use fraction strips to find each sum or difference. Write your answer in simplest form.**

1. $\frac{1}{4} + \frac{1}{8}$    2. $\frac{5}{6} - \frac{2}{3}$    3. $\frac{3}{4} - \frac{1}{3}$    4. $\frac{3}{5} + \frac{3}{10}$

_____    _____    _____    _____

5. $\frac{3}{4} + \frac{1}{6}$    6. $\frac{1}{2} + \frac{3}{8}$    7. $\frac{2}{3} - \frac{1}{6}$    8. $\frac{1}{3} - \frac{1}{4}$

_____    _____    _____    _____

Name _____ Date _____ Class _____

## Challenge
**LESSON 5-2**
*Egyptian Fractions*

Did you know that ancient Egyptians used fractions 5,000 years ago? Some of their fractions were like the ones we use today. However, the Egyptians only used **unit fractions,** or fractions with a numerator of 1. All other fractions had to be written as a sum of unit fractions. And no sum could repeat the same unit fraction! For example, the Egyptians would write $\frac{3}{4}$ as $\frac{1}{2} + \frac{1}{4}$. They would not write $\frac{1}{4} + \frac{1}{4} + \frac{1}{4}$.

**Ancient Egyptians did not have paper. They recorded their math work on papyrus, or thin strips of dried plants. Study the Egyptian fractions recorded on the papyrus scrolls below. Then write each fraction the way we do today.**

1. $\frac{1}{2} + \frac{1}{6}$

2. $\frac{1}{2} + \frac{1}{4} + \frac{1}{20}$

3. $\frac{1}{2} + \frac{1}{3}$

4. $\frac{1}{4} + \frac{1}{8}$

5. $\frac{1}{2} + \frac{1}{3} + \frac{1}{15}$

Name _____ Date _____ Class _____

## LESSON 5-2 Problem Solving
### Adding and Subtracting with Unlike Denominators

Use the circle graph to answer the questions. Write each answer in simplest form.

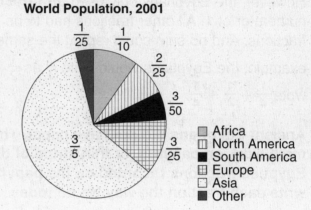

World Population, 2001

1. On which two continents do most people live? How much of the total population do they make up together?

    _____

2. How much of the world's population live in either North America or South America?

    _____

3. How much more of the world's total population lives in Asia than in Africa?

    _____

Circle the letter of the correct answer.

4. How much of Earth's total population do people in Asia and Africa make up all together?

    A $\frac{3}{10}$ of the population

    B $\frac{2}{5}$ of the population

    C $\frac{7}{10}$ of the population

    D $\frac{7}{5}$ of the population

5. What is the difference between North America's part of the total population and Africa's part?

    F Africa has $\frac{1}{50}$ more.

    G Africa has $\frac{1}{50}$ less.

    H Africa has $\frac{9}{50}$ more.

    J Africa has $\frac{9}{50}$ less.

6. How much more of the population lives in Europe than in North America?

    A $\frac{1}{25}$ of the population

    B $\frac{1}{5}$ of the population

    C $\frac{1}{15}$ of the population

    D $\frac{1}{10}$ of the population

7. How much of the world's population lives in North America and Europe?

    F $\frac{1}{25}$ of the population

    G $\frac{1}{15}$ of the population

    H $\frac{1}{5}$ of the population

    J $\frac{1}{20}$ of the population

Name _____ Date _____ Class _____

## LESSON 5-2 Reading Strategies
### Use Fraction Bars

You can use fraction bars to show $\frac{1}{2}$ and $\frac{1}{3}$.

| $\frac{1}{2}$ | $\frac{1}{3}$ |

These fractions have denominators that are different. They are called **unlike fractions**.
To add or subtract unlike fractions, the denominators must be the same. They must have a **common denominator**.

| $\frac{1}{6}$ | $\frac{1}{6}$ | $\frac{1}{6}$ | $\frac{1}{6}$ | $\frac{1}{6}$ |

The common denominator for $\frac{1}{2}$ and $\frac{1}{3}$ is 6.

To get a common denominator for two fractions, multiply the denominators.
halves • thirds = sixths, or 2 • 3 = 6

1. What are unlike fractions?

   _____

2. If you want to add or subtract unlike fractions, what do you need to do?

   _____

3. How do you get a common denominator for $\frac{1}{2}$ and $\frac{1}{3}$?

   _____

4. How many sixths are in one-half?

   _____

5. How many sixths are in one-third?

   _____

6. What is the sum of one-half and one-third?

   _____

7. What is the difference between one-half and one-third? _____

Name _____ Date _____ Class _____

## LESSON 5-2
# Puzzles, Twisters & Teasers
### The Truth of the Matter

Decide whether each statement is true or false. Circle your answer.

Use the letters of your true answers and rearrange them to answer the question.

1. one-third plus one-seventh is greater than one-half
   - **M** true
   - **E** false

2. one-fourth plus one-ninth is greater than one-third
   - **C** true
   - **P** false

3. four-fifths minus one-third is less than one-half
   - **D** true
   - **N** false

4. three-fourths minus three tenths equals one-half
   - **A** true
   - **S** false

5. seven-twelfths plus three-eighths is greater than one
   - **J** true
   - **T** false

6. three-fourths minus three-twelfths equals one-half
   - **L** true
   - **B** false

What must you find in order to add or subtract unlike fractions?

The ___ ___ ___

Name _____ Date _____ Class _____

## LESSON 5-3 Practice A
### Adding and Subtracting Mixed Numbers

**Estimate each sum or difference to the nearest whole number.**

1. $2\frac{1}{5} + 1\frac{1}{4}$

2. $3\frac{1}{6} + 1\frac{4}{5}$

3. $4\frac{1}{2} - 2\frac{1}{8}$

4. $1\frac{1}{2} + 2\frac{3}{4}$

5. $2\frac{2}{3} - 1\frac{5}{6}$

6. $1\frac{1}{7} - 1\frac{1}{8}$

**Find each sum or difference. Write the answer in simplest form.**

7. $1\frac{1}{2} + 3\frac{1}{4}$

8. $10\frac{3}{5} - 8\frac{1}{10}$

9. $3\frac{1}{4} + 2\frac{5}{6}$

10. $3\frac{2}{6} - 1\frac{1}{3}$

11. $10\frac{2}{3} - 9\frac{1}{4}$

12. $4\frac{2}{15} + 1\frac{1}{5}$

13. $8\frac{1}{2} + 2\frac{1}{3}$

14. $12\frac{1}{2} - 10\frac{1}{8}$

15. $7\frac{1}{8} + 1\frac{1}{6}$

16. $2\frac{7}{12} + 1\frac{1}{8}$

17. $4\frac{1}{6} - 1\frac{1}{9}$

18. $3\frac{1}{7} + 3\frac{1}{3}$

19. $1\frac{2}{3} - 1\frac{1}{2}$

20. $5\frac{2}{5} + 1\frac{1}{2}$

21. $2\frac{1}{3} + 2\frac{1}{5}$

22. Jack babysat for $4\frac{1}{4}$ hours on Friday night. He babysat for $3\frac{2}{3}$ hours on Saturday night. How many hours did he babysit in all?

23. Bonita planted an oak tree and an elm tree in her backyard. Three years later, the oak tree was $5\frac{1}{6}$ feet tall, and the elm tree was $7\frac{1}{2}$ feet tall. How much taller was the elm tree?

Holt Mathematics

## LESSON 5-3 Practice B
### Adding and Subtracting Mixed Numbers

Find each sum or difference. Write the answer in simplest form.

1. $4\frac{3}{8} + 5\frac{1}{4}$

2. $11\frac{2}{5} - 8\frac{1}{3}$

3. $7\frac{1}{3} + 3\frac{2}{9}$

_____   _____   _____

4. $22\frac{5}{6} - 17\frac{1}{4}$

5. $32\frac{4}{7} - 14\frac{1}{3}$

6. $12\frac{1}{4} + 5\frac{1}{12}$

_____   _____   _____

7. $29\frac{1}{3} - 14\frac{1}{6}$

8. $5\frac{3}{4} - 1\frac{7}{11}$

9. $21\frac{1}{6} + 1\frac{3}{8}$

_____   _____   _____

10. $15\frac{7}{12} - 14\frac{3}{8}$

11. $5\frac{6}{15} + 4\frac{3}{10}$

12. $25\frac{1}{7} + 25\frac{2}{5}$

_____   _____   _____

13. $3\frac{2}{5} + 1\frac{1}{3}$

14. $1\frac{2}{5} - 1\frac{2}{10}$

15. $3\frac{3}{5} - 2\frac{1}{2}$

_____   _____   _____

16. $6\frac{3}{4} - 3\frac{3}{10}$

17. $4\frac{4}{5} + 2\frac{1}{10}$

18. $32\frac{1}{2} + 5\frac{1}{3}$

_____   _____   _____

19. Donald is making a party mix. He bought $2\frac{1}{4}$ pounds of pecans and $3\frac{1}{5}$ pounds of walnuts. How many pounds of nuts did Donald buy in all? _____

20. Mrs. Watson's cookie recipe calls for $3\frac{4}{7}$ cups of sugar. Mr. Clark's cookie recipe calls for $4\frac{2}{3}$ cups of sugar. How much more sugar does Mr. Clark's recipe use? _____

21. Tasha's cat weighs $15\frac{5}{12}$ lb. Naomi's cat weighs $11\frac{1}{3}$ lb. Can they bring both of their cats to the vet in a carrier that can hold up to 27 pounds? Explain.

_____

Holt Mathematics

Name _____ Date _____ Class _____

## LESSON 5-3 Practice C
### Adding and Subtracting Mixed Numbers

**Add or subtract. Write each answer in simplest form.**

1. $8\frac{1}{10} + 2\frac{2}{25}$

2. $3\frac{1}{2} - 1\frac{10}{17}$

3. $2\frac{1}{6} - 1\frac{1}{16}$

4. $6\frac{5}{6} + 14\frac{1}{9}$

5. $11\frac{7}{8} - 5\frac{3}{12}$

6. $4\frac{2}{13} + 3\frac{1}{2}$

7. $2\frac{1}{12} - 1\frac{1}{14}$

8. $5\frac{7}{16} - 4\frac{2}{5}$

9. $1\frac{1}{8} + 1\frac{1}{9}$

**Evaluate. Write each answer as a fraction in simplest form.**

10. $17\frac{2}{7} + 1.6$

11. $8\frac{1}{5} + 1.5$

12. $19\frac{9}{25} - 6.3$

13. $23\frac{9}{10} - 18.7$

14. $11.42 + \frac{1}{25}$

15. $12\frac{17}{20} - 4.05$

16. Mattie's camping gear and food weighed $24\frac{7}{15}$ pounds at the beginning of the weekend. His food weighed $10\frac{2}{9}$ pounds. After Mattie ate all of his food, how much did his gear weigh? _____

17. Cindy and Tara are saving money to buy a present for their teacher that costs $37.50. So far, Cindy has saved $15\frac{1}{4}$ dollars, and Tara has saved $11\frac{4}{5}$ dollars. How much more do they need to buy the present? _____

18. Terry rode her bike for $15\frac{1}{6}$ miles last week. This week she rode her bike for $21\frac{1}{2}$ miles. How many miles did Terry ride her bike during these two weeks? How many more miles did she ride her bike during the second week? _____

Name _____ Date _____ Class _____

## LESSON 5-3 Reteach
### Adding and Subtracting Mixed Numbers

You can use what you know about improper fractions to add and subtract mixed numbers.

To find the sum or difference of mixed numbers, first write the mixed numbers as improper fractions.

**A.** $3\frac{1}{4} + 2\frac{1}{3}$

$= \frac{13}{4} + \frac{7}{3}$

**B.** $4\frac{1}{2} - 2\frac{2}{3}$

$= \frac{9}{2} - \frac{8}{3}$

Next, find equivalent fractions with a least common denominator.

$\frac{13}{4} + \frac{7}{3}$

$= \frac{39}{12} + \frac{28}{12}$

$\frac{9}{2} - \frac{8}{3}$

$= \frac{27}{6} - \frac{16}{6}$

Then add or subtract the like fractions.

$\frac{39}{12} + \frac{28}{12}$

$= \frac{67}{12}$

$\frac{27}{6} - \frac{16}{6}$

$= \frac{11}{6}$

Write the answer as a mixed number in simplest form.

$\frac{67}{12}$

$= 5\frac{7}{12}$

$\frac{11}{6}$

$= 1\frac{5}{6}$

So, $3\frac{1}{4} + 2\frac{1}{3} = 5\frac{7}{12}$.

So, $4\frac{1}{2} - 2\frac{2}{3} = 1\frac{5}{6}$.

**Find each sum or difference. Write your answer in simplest form.**

1. $1\frac{1}{4} + 1\frac{1}{2}$

$= \frac{\phantom{0}}{4} + \frac{\phantom{0}}{2}$

$= \frac{\phantom{0}}{4} + \frac{\phantom{0}}{4}$

_____

2. $3\frac{1}{6} + 1\frac{2}{3}$

$= \frac{\phantom{0}}{6} + \frac{\phantom{0}}{3}$

$= \frac{\phantom{0}}{6} + \frac{\phantom{0}}{6}$

_____

3. $2\frac{1}{8} + 4\frac{1}{2}$

$= \frac{\phantom{0}}{8} + \frac{\phantom{0}}{2}$

$= \frac{\phantom{0}}{8} + \frac{\phantom{0}}{8}$

_____

4. $4\frac{1}{3} + 1\frac{1}{2}$

$= \frac{\phantom{0}}{3} + \frac{\phantom{0}}{2}$

$= \frac{\phantom{0}}{6} + \frac{\phantom{0}}{6}$

_____

5. $2\frac{3}{5} + 1\frac{1}{10}$

6. $3\frac{1}{6} + 1\frac{1}{12}$

7. $2\frac{5}{8} - 1\frac{1}{4}$

8. $5\frac{2}{3} - 2\frac{1}{4}$

Name _____ Date _____ Class _____

# LESSON 5-3 Challenge
## Maximum Snakes

The bar graph below shows the maximum lengths for the longest snakes in the world. Use the graph to find how much each of the snakes in the City Zoo is below its maximum length.

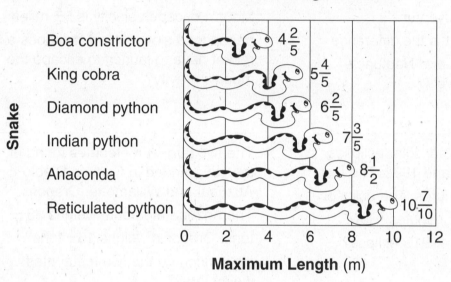

**World's Longest Snakes**

Boa constrictor $4\frac{2}{5}$
King cobra $5\frac{4}{5}$
Diamond python $6\frac{2}{5}$
Indian python $7\frac{3}{5}$
Anaconda $8\frac{1}{2}$
Reticulated python $10\frac{7}{10}$

Maximum Length (m)

**Snakes in the City Zoo**

| Snake | Length (in meters) | Difference from Maximum Length |
|---|---|---|
| Kevin (king cobra) | $3\frac{1}{2}$ | |
| Annie (anaconda) | $5\frac{1}{3}$ | |
| Bob (boa constrictor) | $3\frac{1}{4}$ | |
| Ivy (Indian python) | $4\frac{2}{7}$ | |
| Reggie (reticulated python) | $8\frac{3}{5}$ | |
| Diana (diamond python) | $4\frac{3}{8}$ | |

Name _____ Date _____ Class _____

## Problem Solving
### LESSON 5-3 Adding and Subtracting Mixed Numbers

**Write the correct answer in simplest form.**

1. Of the planets in our solar system, Jupiter and Neptune have the greatest surface gravity. Jupiter's gravitational pull is $2\frac{16}{25}$ stronger than Earth's, and Neptune's is $1\frac{1}{5}$ stronger. What is the difference between Jupiter's and Neptune's surface gravity levels?

2. Escape velocity is the speed a rocket must attain to overcome a planet's gravitational pull. Earth's escape velocity is $6\frac{9}{10}$ miles per second! The Moon's escape velocity is $5\frac{2}{5}$ miles per second slower. How fast does a rocket have to launch to escape the moon's gravity?

3. The two longest total solar eclipses occurred in 1991 and 1992. The first one lasted $6\frac{5}{6}$ minutes. The eclipse of 1992 lasted $5\frac{1}{3}$ minutes. How much longer was 1991's eclipse?

4. The two largest meteorites found in the U.S. landed in Canyon Diablo, Arizona, and Willamette, Oregon. The Arizona meteorite weighs $33\frac{1}{10}$ tons! Oregon's weighs $16\frac{1}{2}$ tons. How much do the two meteorites weigh in all?

**Circle the letter of the correct answer.**

5. Not including the Sun, Proxima Centauri is the closest star to Earth. It is $4\frac{11}{50}$ light years away! The next closest star is Alpha Centauri. It is $\frac{13}{100}$ light years farther than Proxima. How far is Alpha Centauri from Earth?

   A $4\frac{7}{20}$ light years

   B $4\frac{13}{100}$ light years

   C $4\frac{6}{25}$ light years

   D $4\frac{1}{50}$ light years

6. It takes about $5\frac{1}{3}$ minutes for light from the Sun to reach Earth. The Moon is closer to Earth, so its light reaches Earth faster—about $5\frac{19}{60}$ minutes faster than from the Sun. How long does light from the Moon take to reach Earth?

   F $\frac{3}{10}$ of a minute

   G $\frac{1}{60}$ of a minute

   H $\frac{1}{3}$ of a minute

   J $\frac{4}{15}$ of a minute

Name _____ Date _____ Class _____

## LESSON 5-3 Reading Strategies
### Summarize

$\frac{2}{3} + \frac{1}{4}$

To add $\frac{2}{3} + \frac{1}{4}$, the denominators must be the same.

**First,** multiply the denominators to get a common denominator.

$3 \times 4 = 12$

1. How do you get a common denominator for two fractions?

   _____

2. What is the common denominator of $\frac{2}{3}$ and $\frac{1}{4}$? _____

**Next,** change both fractions into equivalent fractions with 12 as the denominator.

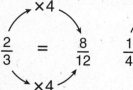 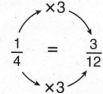

**Finally,** add the fractions together.

**Look at the drawing above to answer each question.**

3. What equivalent fraction is formed when $\frac{2}{3}$ is changed into twelfths?

   _____

4. What equivalent fraction is formed when $\frac{1}{4}$ is changed into twelfths?

   _____

5. What is $\frac{8}{12} + \frac{3}{12}$? _____

6. Write how you would add two fractions that have different denominators.

   _____
   _____

Name _____  Date _____  Class _____

# Puzzles, Twisters & Teasers
## LESSON 5-3 Work Before Play

**Down**

1. The whole number part of the answer: $7\frac{5}{6} - 2\frac{1}{2}$.

2. When you have ____ denominators,

4. you must find the ____.

6. The common denominator of $2\frac{4}{5} + 3\frac{1}{2}$.

**Across**

1. Solve for $n$. $\frac{n}{3} + 2\frac{1}{3} = 3\frac{2}{3}$

3. $\frac{7}{3}$ is the ____ of $\frac{3}{7}$.

5. The fractional part of the answer: $7\frac{5}{6} - 2\frac{1}{2}$

7. The numerator of $\frac{7}{8}$.

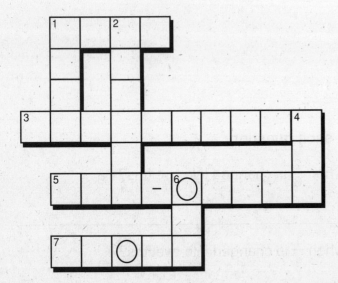

The circled letters in the crossword will give you the answer.

**What two letters can keep you from doing your homework?** ___ ___

# Practice A
## 5-4 Regrouping to Subtract Mixed Numbers

**Regroup each mixed number by regrouping a 1 from the whole number.**

1. $1\frac{1}{4}$

2. $8\frac{5}{12}$

3. $4\frac{5}{9}$

4. $2\frac{1}{3}$

5. $7\frac{1}{9}$

6. $10\frac{3}{7}$

**Subtract. Write each answer in simplest form.**

7. $2 - \frac{2}{3}$

8. $1 - \frac{1}{4}$

9. $5\frac{1}{4} - 3\frac{1}{2}$

10. $2\frac{1}{3} - 1\frac{5}{6}$

11. $1\frac{4}{9} - \frac{2}{3}$

12. $2\frac{1}{4} - 1\frac{7}{8}$

13. $5\frac{3}{10} - 1\frac{4}{5}$

14. $2\frac{1}{4} - \frac{11}{16}$

15. $3\frac{1}{3} - 2\frac{4}{5}$

16. At the pie-eating contest, Dina ate $3\frac{1}{3}$ pies. Mason ate $2\frac{5}{6}$ pies. How much more pie did Dina eat than Mason?

17. When Latoya bought her angel fish, it was $1\frac{1}{2}$ inches long. Now it is $2\frac{1}{3}$ inches long. How much did her angel fish grow?

Name _____ Date _____ Class _____

## LESSON 5-4 Practice B
### Regrouping to Subtract Mixed Numbers

**Subtract. Write each answer in simplest form.**

1. $4 - 2\frac{3}{8}$

2. $5\frac{1}{6} - 2\frac{2}{3}$

3. $14 - 8\frac{2}{9}$

4. $19\frac{1}{7} - 5\frac{1}{3}$

5. $7\frac{1}{4} - 3\frac{5}{8}$

6. $10\frac{1}{5} - 5\frac{7}{10}$

7. $1\frac{1}{6} - \frac{7}{9}$

8. $9\frac{1}{4} - 1\frac{7}{16}$

9. $6\frac{1}{5} - 3\frac{1}{4}$

**Evaluate each expression for $a = 1\frac{1}{2}$, $b = 2\frac{1}{3}$, $c = \frac{1}{4}$, and $d = 3$. Write the answer in simplest form.**

10. $b - a$

11. $a - c$

12. $b - c$

13. $d - a$

14. $d - b$

15. $d - c$

16. Tim had 6 feet of wrapping paper for Kylie's birthday present. He used $3\frac{3}{8}$ feet of the paper to wrap her gift. How much paper did Tim have left? _____

17. At his last doctor's visit, Pablo was $60\frac{1}{2}$ inches tall. At today's visit, he measured $61\frac{1}{6}$ inches. How much did Pablo grow between visits? _____

18. Yesterday, Danielle rode her bike for $5\frac{1}{2}$ miles. Today, she rode her bike for $6\frac{1}{4}$ miles. How much farther did Danielle ride her bike today? _____

# Practice C
## 5-4 Regrouping to Subtract Mixed Numbers

**Subtract. Write each answer in simplest form.**

1. $7 - 3\frac{11}{12}$

2. $8\frac{4}{13} - 1\frac{19}{26}$

3. $14\frac{5}{12} - 3\frac{7}{8}$

4. $5\frac{1}{7} - 2\frac{2}{3}$

5. $19\frac{1}{12} - 4\frac{4}{9}$

6. $19\frac{3}{5} - 6\frac{5}{7}$

7. $17\frac{1}{14} - 8\frac{7}{8}$

8. $14\frac{2}{7} - 11\frac{3}{4}$

9. $22\frac{3}{11} - 2\frac{9}{10}$

**Evaluate each expression for $a = 6\frac{3}{8}$, $b = 5\frac{1}{6}$, $c = 7\frac{1}{12}$, and $d = 10$. Write the answer in simplest form.**

10. $a - b$

11. $c - a$

12. $c - b$

13. $d - a$

14. $d - b$

15. $d - c$

16. Annie bought $21\frac{2}{5}$ pounds of clay. She used $15\frac{5}{6}$ pounds of the clay to make a vase, and $1\frac{4}{5}$ pounds to make a coaster. How much clay does she have left? _____

17. In January, a chef bought $15\frac{1}{8}$ pounds of ground beef. In February, he bought $3\frac{4}{5}$ pounds less. Then in March he bought $1\frac{19}{20}$ pounds less than in February. How many pounds of ground beef did the chef buy in March? _____

18. John is training for the triathlon. He wants to cover a distance of $16\frac{1}{4}$ miles today. If he runs for $6\frac{1}{2}$ miles and rides his bike for $7\frac{4}{5}$ miles, how far does he have to swim? _____

Name _____ Date _____ Class _____

## LESSON 5-4 Reteach
### Regrouping to Subtract Mixed Numbers

You can use fraction strips to regroup to subtract mixed numbers.

To find $3\frac{1}{4} - 1\frac{3}{4}$, first model the first mixed number in the expression.

| 1 | 1 | 1 | $\frac{1}{4}$ |

There are not enough $\frac{1}{4}$ pieces to subtract, so you have to regroup.
Trade one one-whole strip for four $\frac{1}{4}$ pieces, because $\frac{4}{4} = 1$.

| 1 | 1 | $\frac{1}{4}$ $\frac{1}{4}$ $\frac{1}{4}$ $\frac{1}{4}$ | $\frac{1}{4}$ |

Now there are enough $\frac{1}{4}$ pieces to subtract. Take away $1\frac{3}{4}$.

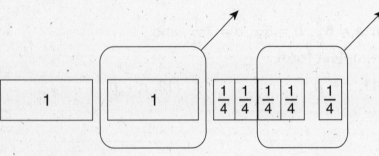

The remaining pieces represent the difference. Write the difference in simplest form.

$3\frac{1}{4} - 1\frac{3}{4} = 1\frac{2}{4} = 1\frac{1}{2}$

**Use fraction strips to find each difference. Write your answer in simplest form.**

1. $3\frac{1}{4} - 2\frac{3}{4}$      2. $3\frac{1}{6} - 1\frac{5}{6}$      3. $4\frac{3}{8} - 1\frac{7}{8}$      4. $3\frac{1}{3} - 2\frac{2}{3}$

_____     _____     _____     _____

5. $5\frac{5}{12} - 2\frac{7}{12}$      6. $3\frac{3}{10} - 1\frac{9}{10}$      7. $5\frac{1}{8} - 1\frac{5}{8}$      8. $4 - 1\frac{1}{3}$

_____     _____     _____     _____

9. $3\frac{1}{8} - 1\frac{3}{8}$      10. $2\frac{1}{8} - 1\frac{7}{8}$      11. $3 - 1\frac{1}{4}$      12. $6\frac{3}{8} - 2\frac{5}{8}$

_____     _____     _____     _____

# Challenge
## LESSON 5-4  Popular First Names

What are the most popular first names in the United States?

**Regroup fractions or mixed numbers to solve each problem below. Write your answers in simplest form. Then, in the box at the bottom of the page, write each problem's letter in the blanks above its solution. When you have solved all the problems, you will have found the answer to the question.**

$8\frac{7}{12} - 7\frac{3}{4}$     __$\frac{5}{6}$__    A

$9\frac{1}{8} - 8\frac{3}{4}$     __$\frac{3}{8}$__    E

$10\frac{1}{3} - 9\frac{2}{3}$     __$\frac{2}{3}$__    J

$6\frac{1}{2} - 5\frac{4}{5}$     __$\frac{7}{10}$__    M

$5\frac{1}{5} - 4\frac{4}{5}$     __$\frac{2}{5}$__    R

$7\frac{2}{9} - 6\frac{2}{3}$     __$\frac{5}{9}$__    S

$12\frac{2}{5} - 11\frac{1}{2}$     __$\frac{9}{10}$__    Y

---

**#1 Name For American Men:** __J__ __A__ __M__ __E__ __S__
                                       $\frac{2}{3}$   $\frac{5}{6}$   $\frac{7}{10}$   $\frac{3}{8}$   $\frac{5}{9}$

**#1 Name For American Women:** __M__ __A__ __R__ __Y__
                                               $\frac{7}{10}$   $\frac{5}{6}$   $\frac{2}{5}$   $\frac{9}{10}$

Name _____ Date _____ Class _____

## LESSON 5-4 Problem Solving
### Regrouping to Subtract Mixed Numbers

**Write the correct answer in simplest form.**

1. The average person in the United States eats $6\frac{13}{16}$ pounds of potato chips each year. The average person in Ireland eats $5\frac{15}{16}$ pounds. How much more potato chips do Americans eat a year than people in Ireland?

2. The average person in the United States eats $270\frac{1}{16}$ pounds of meat each year. The average person in Australia eats $238\frac{1}{2}$ pounds. How much more meat do Americans eat a year than people in Australia?

3. The average Americans eats $24\frac{1}{2}$ pounds of ice cream every year. The average person in Israel eats $15\frac{4}{5}$ pounds. How much more ice cream do Americans eat each year?

4. People in Switzerland eat the most chocolate—26 pounds a year per person. Most Americans eat $12\frac{9}{16}$ pounds each year. How much more chocolate do the Swiss eat?

5. The average person in the United States chews $1\frac{9}{16}$ pounds of gum each year. The average person in Japan chews $\frac{7}{8}$ pound. How much more gum do Americans chew?

6. Norwegians eat the most frozen foods—$78\frac{1}{2}$ pounds per person each year. Most Americans eat $35\frac{15}{16}$ pounds. How much more frozen foods do people in Norway eat?

**Circle the letter of the correct answer.**

7. Most people around the world eat $41\frac{7}{8}$ pounds of sugar each year. Most Americans eat $66\frac{3}{4}$ pounds. How much more sugar do Americans eat than the world's average?

   A $25\frac{7}{8}$ pounds more

   B $25\frac{1}{8}$ pounds more

   C $24\frac{7}{8}$ pounds more

   D $24\frac{1}{8}$ pounds more

8. The average person eats 208 pounds of vegetables and $125\frac{5}{8}$ pounds of fruit each year. How much more vegetables do most people eat than fruit?

   F $83\frac{5}{8}$ pounds more

   G $82\frac{3}{8}$ pounds more

   H $123\frac{5}{8}$ pounds more

   J $83\frac{3}{8}$ pounds more

Name _____ Date _____ Class _____

## LESSON 5-4 Reading Strategies
### Compare and Contrast

When you subtract whole numbers, you often need to regroup a number before you can subtract.

$$\begin{array}{r} \overset{7\ 13}{8\cancel{3}} \\ -17 \\ \hline \end{array}$$

In the above example, there were only 3 ones—not enough to subtract 7. A ten was regrouped as ten ones. The ten ones were added to the three ones to make 13 ones. Now there are enough ones to subtract 7.

You can compare regrouping fractions to regrouping whole numbers.

$$\begin{array}{r} 3\frac{1}{8} \\ -1\frac{3}{8} \\ \hline \end{array}$$

Look at the fractions first. There aren't enough eighths to subtract $\frac{3}{8}$ from $\frac{1}{8}$. Regrouping fractions is different from regrouping whole numbers, because you regroup a whole number as a fraction. You can regroup 1 as a fraction with the same numerator and denominator.

$1 = \frac{2}{2}$     $1 = \frac{5}{5}$     $1 = \frac{8}{8}$

$$\begin{array}{r} 2\frac{9}{8} \\ \cancel{3}\frac{1}{8} \\ -1\frac{3}{8} \\ \hline \end{array}$$

Take one from three and regroup as it $\frac{8}{8}$. Combine $\frac{8}{8}$ with $\frac{1}{8}$ to make $\frac{9}{8}$. Now there are enough eighths to subtract.

1. What is $\frac{9}{8} - \frac{3}{8}$? _____

2. What is the same about subtracting whole numbers and subtracting fractions?

_____

3. What is different about subtracting whole numbers and subtracting fractions?

_____

_____

Copyright © by Holt, Rinehart and Winston.
All rights reserved.

Holt Mathematics

# Puzzles, Twisters & Teasers
## 5-4 Subtraction Chains

Start with the first number in the chain. Subtract the next number, and the next, and the next. If, however, the next number to be subtracted is larger than your current answer, end the chain. Circle the last number you were able to subtract in that chain.

Example chain: | 7 | 2 | 4 | 3 |   7 − 2 = 5. 5 − 4 = 1. Stop now, because 3 is larger than your current answer. Circle 4, the last number you were able to subtract.

1.
| $1\frac{2}{3}$ | $1\frac{1}{3}$ | $\frac{2}{3}$ | $3\frac{1}{3}$ |
|---|---|---|---|
| O | H | N | J |

2.
| $3\frac{1}{7}$ | $1\frac{5}{7}$ | $\frac{5}{7}$ | $\frac{4}{7}$ |
|---|---|---|---|
| U | I | M | E |

3.
| $5\frac{1}{5}$ | $3\frac{3}{5}$ | $\frac{4}{5}$ | $1\frac{1}{5}$ |
|---|---|---|---|
| C | Y | T | H |

4.
| $4\frac{1}{8}$ | $1\frac{5}{8}$ | $2\frac{5}{8}$ | $\frac{1}{8}$ |
|---|---|---|---|
| B | A | W | E |

5.
| $3\frac{3}{12}$ | $\frac{8}{12}$ | $\frac{9}{12}$ | $\frac{7}{12}$ |
|---|---|---|---|
| V | F | R | S |

Now you are ready to solve the riddle. Place the letters for the circled numbers in the numbered spaces and you will have your answer!

In a contest at a local restaurant, the restaurant owner hung two sirloins from the ceiling. Anyone who could jump up and get one won a free dinner. A customer came in, but when he was asked if he would like to try, he responded: "No thanks,

___ ___ ___ ___ ___ ___ ___ K ___ R ___
 3    1    2    5    3    2    4        5    4        2

___ O O ___ I G ___
 3             1              1
"

# Practice A
## LESSON 5-5 Solving Fraction Equations: Addition and Subtraction

Solve each equation. Write the solution in simplest form.

1. $k + 1\frac{1}{2} = 3$

2. $m - 2\frac{1}{3} = 1\frac{1}{2}$

3. $1\frac{1}{4} - \frac{2}{3} = p$

_____  _____  _____

4. $n + 3\frac{7}{8} = 5\frac{1}{8}$

5. $3\frac{1}{3} = y - 1\frac{1}{6}$

6. $2\frac{1}{5} + d = 3\frac{1}{2}$

_____  _____  _____

7. $2\frac{1}{7} + q = 4\frac{3}{14}$

8. $z - 1\frac{2}{5} = 1\frac{7}{10}$

9. $f + \frac{2}{3} = 1\frac{1}{9}$

_____  _____  _____

10. $b = 1\frac{5}{8} - \frac{3}{4}$

11. $t + 1\frac{1}{5} = 3\frac{3}{10}$

12. $3\frac{1}{2} + w = 5\frac{7}{12}$

_____  _____  _____

13. $c - 8\frac{1}{5} = 10\frac{3}{10}$

14. $h + \frac{1}{3} = 2\frac{1}{6}$

15. $1\frac{5}{9} = g - 3\frac{5}{18}$

_____  _____  _____

16. Joey beat Frank in the swim race by $2\frac{1}{10}$ minutes. Frank's time was $8\frac{3}{5}$ minutes. What was Joey's time in the race?

_____

17. Sabrina bought 8 gallons of paint. After she painted her kitchen, she had $4\frac{1}{6}$ gallons left over. How much paint did Sabrina use in her kitchen?

_____

Name _____ Date _____ Class _____

## LESSON 5-5 Practice B
### Solving Fraction Equations: Addition and Subtraction

Solve each equation. Write the solution in simplest form. Check your answers.

1. $k + 3\frac{3}{4} = 5\frac{2}{3} - 1\frac{1}{3}$

2. $a - 2\frac{2}{11} = 2\frac{5}{22} - 1\frac{2}{11}$

3. $2\frac{2}{7} = n - 4\frac{2}{3} - 1\frac{1}{3}$

4. $6\frac{1}{4} = z + 1\frac{5}{8}$

5. $5\frac{1}{4} = x + \frac{7}{16}$

6. $r + 6 = 9\frac{2}{5} - 2\frac{1}{2}$

7. $11\frac{2}{5} = q - 4\frac{2}{7} + 2\frac{1}{7}$

8. $4\frac{2}{5} - 2\frac{1}{2} = p + \frac{3}{10}$

9. $\frac{3}{8} + \frac{1}{6} = c - 4\frac{5}{6}$

10. $2\frac{1}{4} + c = 2\frac{1}{3} + 1\frac{1}{6}$

11. A seamstress raised the hem on Helen's skirt by $1\frac{1}{3}$ inches. The skirt's original length was 16 inches. What is the new length?

12. The bike trail is $5\frac{1}{4}$ miles long. Jessie has already cycled $2\frac{5}{8}$ miles of the trail. How much farther does she need to go to finish the trail?

Holt Mathematics

Name _____ Date _____ Class _____

## LESSON 5-5 Practice C
### Solving Fraction Equations: Addition and Subtraction

Solve each equation. Write the solution in simplest form. Check your answers.

1. $3\frac{1}{5} - 1\frac{3}{10} = p + \frac{5}{6}$

2. $17\frac{5}{6} + \frac{7}{10} = d - 2\frac{5}{12}$

3. $34\frac{1}{6} = x + 6\frac{1}{4} + 12\frac{3}{8}$

4. $a - 2\frac{3}{11} = 19\frac{1}{2} - 16\frac{1}{4}$

5. $f - 4\frac{1}{10} + 15\frac{3}{5} = 29\frac{18}{25}$

6. $\frac{7}{12} + \frac{3}{8} = c - 2\frac{5}{6}$

7. $r + 11\frac{3}{5} = 20\frac{1}{5} - 3\frac{1}{2}$

8. $s + 30\frac{11}{15} = 40\frac{1}{3} - 2\frac{1}{2}$

9. Carol wants each of the curtains she makes to be the same length. She started with two pieces of cloth measuring $6\frac{1}{3}$ feet and $7\frac{3}{4}$ feet. She cut $1\frac{5}{8}$ feet off the $6\frac{1}{3}$-foot piece. How much should she cut from the second piece?

10. Last year it rained $42\frac{1}{6}$ inches in Portland, Maine. It rained $14\frac{5}{8}$ inches in the spring, and $11\frac{1}{24}$ inches in the summer. The city received the same amounts of rain in the fall as in winter. How much did it rain in fall?

## LESSON 5-5 Reteach
### Solving Fraction Equations: Addition and Subtraction

You can write related facts using addition and subtraction.

$3 + 4 = 7 \qquad 7 - 4 = 3$

You can use related facts to solve equations.

**A.** $x + 2\frac{1}{2} = 4$

Think: $4 - 2\frac{1}{2} = x$

$x = 4 - 2\frac{1}{2}$

$x = 3\frac{2}{2} - 2\frac{1}{2}$  Regroup 4 as $3\frac{2}{2}$.

$x = 1\frac{1}{2}$

**B.** $x - 4\frac{1}{3} = 3\frac{1}{2}$

Think: $3\frac{1}{2} + 4\frac{1}{3} = x$

$x = 3\frac{1}{2} + 4\frac{1}{3}$

$x = \frac{7}{2} + \frac{13}{3}$  Write the mixed numbers as improper fractions.

$x = \frac{21}{6} + \frac{26}{6}$  Write the fractions using a common denominator.

$x = \frac{47}{6}$

$x = 7\frac{5}{6}$  Write the sum as a mixed number.

**Use related facts to solve each equation.**

1. $x + 3\frac{1}{3} = 7$

   $x = 7 - 3\frac{1}{3}$

   $x = 6\frac{3}{3} - 3\frac{1}{3}$

   $x = \underline{\qquad}$

2. $x - 2\frac{1}{4} = 4\frac{1}{2}$

   $x = 4\frac{1}{2} + 2\frac{1}{4}$

   $x = \frac{9}{2} + \frac{9}{4}$

   $x = \frac{18}{4} + \frac{9}{4}$

   $x = \underline{\qquad}$

3. $x + \frac{3}{8} = 5\frac{1}{4}$

   $x = 5\frac{1}{4} - \frac{3}{8}$

   $x = \frac{21}{4} - \frac{3}{8}$

   $x = \frac{42}{8} - \frac{3}{8}$

   $x = \underline{\qquad}$

4. $x - \frac{5}{12} = 2\frac{1}{2}$

   $x = 2\frac{1}{2} + \frac{5}{12}$

   $x = \frac{5}{2} + \frac{5}{12}$

   $x = \frac{30}{12} + \frac{5}{12}$

   $x = \underline{\qquad}$

5. $x - 1\frac{3}{4} = 7\frac{1}{2}$

6. $x - 3\frac{2}{3} = 1\frac{1}{3}$

7. $x + 3\frac{1}{2} = 6\frac{1}{4}$

8. $x - 2\frac{2}{5} = 1\frac{3}{10}$

Name _____ Date _____ Class _____

## LESSON 5-5 Challenge
### You Read My Mind!

Here's a trick you can use to amaze your friends and family. Start by asking your friends to think of a number—any number. Pretend you are reading their minds while you write the number 6 on a piece of paper. (Don't show it to them.) Then use the steps below to tell them what to do. The fraction $\frac{2}{5}$ is used as an example choice, but the trick works for any chosen fraction, mixed number, decimal, or whole number.

| Step | Example |
|---|---|
| 1. Double your number. | $\frac{2}{5} + \frac{2}{5} = \frac{4}{5}$ |
| 2. Add 12 to your sum. | $\frac{4}{5} + 12 = 12\frac{4}{5}$ |
| 3. Divide your new sum by 2. | $12\frac{4}{5} \div 2 = 6\frac{2}{5}$ |
| 4. Subtract your chosen number from that quotient. | $6\frac{2}{5} - \frac{2}{5} = 6$ |

Now amaze your friends by showing that you wrote the same number they ended with. No matter what number is chosen, this trick always ends in 6. Equations explain why it works—but don't tell your friends this part.

Let $x$ = the chosen number.

| STEP 1 | STEP 2 | STEP 3 | STEP 4 |
|---|---|---|---|
| ↓ | ↓ | ↓ | ↓ |
| $2x$ | $2x + 12$ | $(2x + 12) \div 2 = x + 6$ | $x + 6 - x = 6$ |

Before you try the trick, practice it on the fractions below. Use the equations for each step and show all your work.

1. Chosen Number: $\frac{7}{9}$

   STEP 1: _____
   
   STEP 2: _____
   
   STEP 3: _____
   
   STEP 4: _____

2. Chosen Number: $3\frac{1}{4}$

   STEP 1: _____
   
   STEP 2: _____
   
   STEP 3: _____
   
   STEP 4: _____

Name _____ Date _____ Class _____

## LESSON 5-5 Problem Solving
### Solving Fraction Equations: Addition and Subtraction

**Write the correct answer in simplest form.**

1. It usually takes Brian $1\frac{1}{2}$ hours to get to work from the time he gets out of bed. His drive to the office takes $\frac{3}{4}$ hour. How much time does he spend getting ready for work?

   _____

2. Before she went to the hairdresser, Sheila's hair was $7\frac{1}{4}$ inches long. When she left the salon, it was $5\frac{1}{2}$ inches long. How much of her hair did Sheila get cut off?

   _____

3. One lap around the gym is $\frac{1}{3}$ mile long. Kim has already run 5 times around. If she wants to run 2 miles total, how much farther does she have to go?

   _____

4. Darius timed his speech at $5\frac{1}{6}$ minutes. His time limit for the speech is $4\frac{1}{2}$ minutes. How much does he need to cut from his speech?

   _____

**Circle the letter of the correct answer.**

5. Mei and Alex bought the same amount of food at the deli. Mei bought $1\frac{1}{4}$ pounds of turkey and $1\frac{1}{3}$ pounds of cheese. Alex bought $1\frac{1}{2}$ pounds of turkey. How much cheese did Alex buy?

   A $1\frac{1}{12}$ pounds  C $1\frac{1}{4}$ pounds

   B $1\frac{1}{6}$ pounds  D $4\frac{1}{12}$ pounds

6. When Lynn got her dog, Max, he weighed $10\frac{1}{2}$ pounds. During the next 6 months, he gained $8\frac{4}{5}$ pounds. At his one-year check-up he had gained another $4\frac{1}{3}$ pounds. How much did Max weigh when he was 1 year old?

   F $22\frac{19}{30}$ pounds  H $23\frac{29}{30}$ pounds

   G $23\frac{19}{30}$ pounds  J $23\frac{49}{50}$ pounds

7. Charlie picked up 2 planks of wood at the hardware store. One is $6\frac{1}{4}$ feet long and the other is $5\frac{5}{8}$ feet long. How much should he cut from the first plank to make them the same length?

   A $\frac{5}{8}$ foot  C $1\frac{3}{8}$ feet

   B $\frac{1}{2}$ foot  D $1\frac{5}{8}$ feet

8. Carmen used $3\frac{3}{4}$ cups of flour to make a cake. She had $\frac{1}{2}$ cup of flour left over. Which equation can you use to find how much flour she had before baking the cake?

   F $x + \frac{1}{2} = 3\frac{3}{4}$  H $3\frac{3}{4} - \frac{1}{2} = x$

   G $x - 3\frac{3}{4} = \frac{1}{2}$  J $3\frac{3}{4} - x = \frac{1}{2}$

Name _____ Date _____ Class _____

## LESSON 5-5 Reading Strategies
### Summarize

The following steps are used to solve addition and subtraction equations with fractions.

$$2\tfrac{1}{3} + m = 5$$

$$2\tfrac{1}{3} - 2\tfrac{1}{3} + m = 5 - 2\tfrac{1}{3}$$ ← **Step 1:** Subtract $2\tfrac{1}{3}$ from both sides of the equation.

$$m = 5 - 2\tfrac{1}{3}$$

$$m = 4\tfrac{3}{3} - 2\tfrac{1}{3}$$ ← **Step 2:** Regroup 5 as $4\tfrac{3}{3}$

$$m = 2\tfrac{2}{3}$$ ← **Step 3:** Subtract fractions. Subtract whole numbers.

**Answer each question.**

1. What is the first step in the example above?

   _____

2. Why was $2\tfrac{1}{3}$ subtracted from both sides of the equation?

   _____

3. What is the second step in the example above?

   _____

**Use this equation to answer the following questions:**

$$x - 3\tfrac{2}{3} = 2\tfrac{2}{3}$$

4. What is the first step to solve the equation?

   _____

5. What is the next step to solve the equation?

   _____

6. Write how you solve equations that involve fractions.

   _____
   _____

## Puzzles, Twisters & Teasers
**LESSON 5-5** *Leap to Success*

Solve the equations below and record the answers. Match the answer to the letter of the variable in the equation the answer goes with.

$R - \frac{4}{5} = \frac{1}{10}$    $R = $ _____

$2\frac{2}{3} + O = 3\frac{1}{8}$    $O = $ _____

$C + 6\frac{7}{8} = 10\frac{1}{5}$    $C = $ _____

$I - \frac{1}{11} = \frac{9}{22}$    $I = $ _____

$N - 2\frac{7}{9} = 1\frac{1}{2}$    $N = $ _____

$8\frac{3}{5} = A + 4\frac{6}{7}$    $A = $ _____

$5\frac{1}{4} + K = 6\frac{5}{12}$    $K = $ _____

$G + \frac{4}{5} = 2\frac{2}{21}$    $G = $ _____

Why were frogs put on the endangered species list?

Because they are always ___ ___ ___ ___ ___ ___ ___ ___ !
$\quad\quad\quad\quad\quad\quad\quad\quad\quad\quad 3\frac{13}{40}\ \ \frac{9}{10}\ \ \frac{11}{24}\ \ 3\frac{26}{35}\ \ 1\frac{1}{6}\ \ \frac{1}{2}\ \ 4\frac{5}{18}\ \ 1\frac{31}{105}$

Name _____ Date _____ Class _____

## LESSON 5-6 Practice A
### Multiplying Fractions Using Repeated Addition

**Multiply. Write each answer in simplest form.**

1. $1 \cdot \frac{1}{3}$

2. $3 \cdot \frac{1}{8}$

3. $7 \cdot \frac{1}{9}$

4. $3 \cdot \frac{1}{4}$

5. $4 \cdot \frac{2}{10}$

6. $3 \cdot \frac{1}{6}$

7. $2 \cdot \frac{2}{5}$

8. $10 \cdot \frac{1}{2}$

9. $5 \cdot \frac{1}{8}$

10. $4 \cdot \frac{1}{6}$

11. $5 \cdot \frac{1}{8}$

12. $3 \cdot \frac{2}{6}$

13. $7 \cdot \frac{1}{11}$

14. $3 \cdot \frac{1}{9}$

15. $5 \cdot \frac{1}{15}$

**Evaluate 2x for each value of x. Write the answer in simplest form.**

16. $x = \frac{1}{4}$

17. $x = \frac{1}{3}$

18. $x = \frac{1}{2}$

19. $x = \frac{1}{6}$

20. $x = \frac{1}{7}$

21. $x = \frac{1}{8}$

22. $x = \frac{2}{3}$

23. $x = \frac{3}{4}$

24. Richie is making 3 quarts of fruit punch for his friends. He must add $\frac{1}{2}$ cup sugar to make each quart of punch. How much sugar will he add?

25. Mrs. Flynn has 20 students in her class. One-fourth of her students purchased lunch tokens. How many of her students purchased tokens?

Name _____ Date _____ Class _____

## LESSON 5-6 Practice B
### Multiplying Fractions Using Repeated Addition

**Multiply. Write each answer in simplest form.**

1. $5 \cdot \frac{1}{10}$　　　　2. $6 \cdot \frac{1}{18}$　　　　3. $4 \cdot \frac{1}{14}$

　_____　　　_____　　　_____

4. $3 \cdot \frac{1}{12}$　　　　5. $2 \cdot \frac{1}{8}$　　　　6. $6 \cdot \frac{1}{10}$

　_____　　　_____　　　_____

7. $3 \cdot \frac{1}{6}$　　　　8. $3 \cdot \frac{5}{12}$　　　　9. $3 \cdot \frac{2}{7}$

　_____　　　_____　　　_____

10. $2 \cdot \frac{3}{8}$　　　　11. $10 \cdot \frac{3}{15}$　　　　12. $8 \cdot \frac{2}{14}$

　_____　　　_____　　　_____

13. $5 \cdot \frac{2}{10}$　　　　14. $4 \cdot \frac{4}{12}$　　　　15. $2 \cdot \frac{13}{20}$

　_____　　　_____　　　_____

**Evaluate 6x for each value of x. Write the answer in simplest form.**

16. $x = \frac{2}{3}$　　17. $x = \frac{2}{8}$　　18. $x = \frac{1}{4}$　　19. $x = \frac{2}{6}$

　_____　_____　_____　_____

20. $x = \frac{2}{7}$　　21. $x = \frac{2}{5}$　　22. $x = \frac{3}{11}$　　23. $x = \frac{5}{12}$

　_____　_____　_____　_____

24. Thomas spends 60 minutes exercising. For $\frac{1}{4}$ of that time, he jumps rope. How many minutes does he spend jumping rope?

25. Kylie made a 4-ounce milk shake. Two-thirds of the milk shake was ice cream. How many ounces of ice cream did Kylie use in the shake?

Copyright © by Holt, Rinehart and Winston.
All rights reserved.

Holt Mathematics

Name _____ Date _____ Class _____

## LESSON 5-6 Practice C
### Multiplying Fractions Using Repeated Addition

**Multiply. Write each answer in simplest form.**

1. $3 \cdot \frac{4}{17}$

2. $2 \cdot \frac{6}{10}$

3. $4 \cdot \frac{3}{4}$

4. $5 \cdot \frac{6}{15}$

5. $3 \cdot \frac{8}{9}$

6. $6 \cdot \frac{3}{14}$

7. $12 \cdot \frac{3}{42}$

8. $6 \cdot \frac{4}{27}$

9. $2 \cdot \frac{16}{20}$

**Evaluate 9x for each value of x. Write the answer in simplest form.**

10. $x = \frac{2}{9}$

11. $x = \frac{3}{36}$

12. $x = \frac{9}{18}$

13. $x = \frac{4}{7}$

14. $x = \frac{5}{12}$

15. $x = \frac{4}{81}$

16. $x = \frac{16}{18}$

17. $x = \frac{27}{50}$

**Evaluate each expression. Write each answer in simplest form.**

18. $9c$ for $c = \frac{2}{3}$

19. $13d$ for $d = \frac{2}{9}$

20. $7n$ for $n = \frac{1}{7}$

**Compare. Write <, >, or =.**

21. $3 \cdot \frac{1}{2}$ ☐ $\frac{4}{7}$

22. $\frac{3}{5}$ ☐ $3 \cdot \frac{1}{5}$

23. $12 \cdot \frac{3}{4}$ ☐ $10$

24. Clair's paycheck this week was $568.00. She put $\frac{1}{4}$ of that amount in her savings account. Then she spent $\frac{1}{2}$ of what was left on rent and $42.60 on groceries. How much money does she have left?

25. A television news program questioned 270 people to see if they voted in the election. Of those questioned, $\frac{2}{15}$ did not vote in the election. How many of those questioned did vote?

Name _____ Date _____ Class _____

## LESSON 5-6 Reteach
### Multiplying Fractions Using Repeated Addition

You can use fraction strips to multiply fractions by whole numbers.

To find $3 \cdot \frac{2}{3}$, first think about the expression in words.

$3 \cdot \frac{2}{3}$ means "3 groups of $\frac{2}{3}$."

Then model the expression.

$\boxed{\frac{1}{3}\,\frac{1}{3}} + \boxed{\frac{1}{3}\,\frac{1}{3}} + \boxed{\frac{1}{3}\,\frac{1}{3}}$

The total number of $\frac{1}{3}$ fraction pieces is 6.

So, $3 \cdot \frac{2}{3} = \frac{2}{3} + \frac{2}{3} + \frac{2}{3} = \frac{6}{3} = 2$ in simplest form.

**Use fraction strips to find each product.**

1. $4 \cdot \frac{1}{8}$  
2. $2 \cdot \frac{2}{5}$  
3. $6 \cdot \frac{1}{8}$  
4. $8 \cdot \frac{1}{4}$

_____   _____   _____   _____

You can also use counters to multiply fractions by whole numbers.

To find $\frac{1}{2} \cdot 12$, first think about the expression in words.

$\frac{1}{2} \cdot 12 = \frac{12}{2}$, which means "12 divided into 2 equal groups."

Then model the expression.

The number of counters in each group is the product.

$\frac{1}{2} \cdot 12 = 6.$

**Use counters to find each product.**

5. $\frac{1}{3} \cdot 15$  
6. $\frac{1}{8} \cdot 24$  
7. $\frac{1}{4} \cdot 16$  
8. $\frac{1}{12} \cdot 24$

Name _____ Date _____ Class _____

## Challenge
**LESSON 5-6**
*Slowpoke Race*

The animals shown below are some of the slowest creatures on Earth. Use their given average speeds to find how far they will travel in the times marked along their racetracks.

Which of these slowpokes traveled the farthest? _____

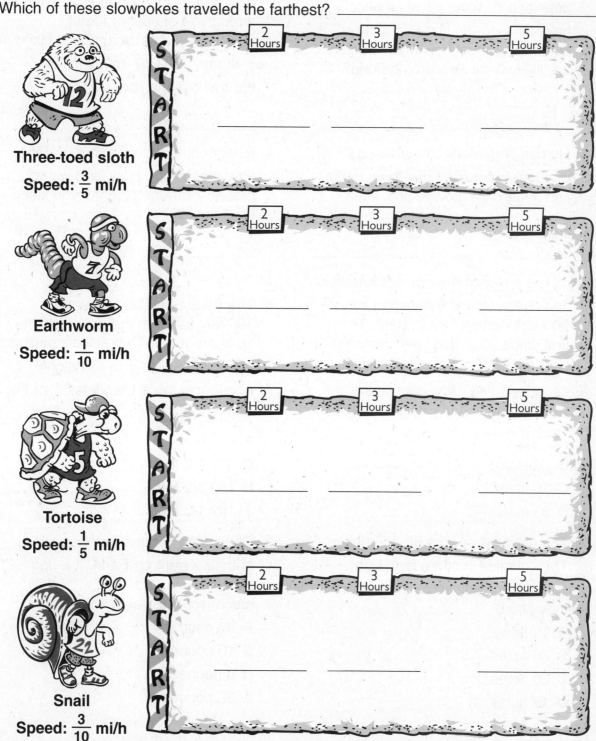

**Three-toed sloth**
Speed: $\frac{3}{5}$ mi/h

**Earthworm**
Speed: $\frac{1}{10}$ mi/h

**Tortoise**
Speed: $\frac{1}{5}$ mi/h

**Snail**
Speed: $\frac{3}{10}$ mi/h

Name _____ Date _____ Class _____

## Problem Solving
### LESSON 5-6 Multiplying Fractions Using Repeated Addition

**Write the answers in simplest form.**

1. Did you know that some people have more bones than the rest of the population? About $\frac{1}{20}$ of all people have an extra rib bone. In a crowd of 60 people, about how many people are likely have an extra rib bone?

2. The Appalachian National Scenic Trail is the longest marked walking path in the United States. It extends through 14 states for about 2,000 miles. Last year, Carla hiked $\frac{1}{5}$ of the trail. How many miles of the trail did she hike?

3. Human fingernails can grow up to $\frac{1}{10}$ of a millimeter each day. How much can fingernails grow in one week?

4. Most people dream about $\frac{1}{4}$ of the time they sleep. How long will you probably dream tonight if you sleep for 8 hours?

**Circle the letter of the correct answer.**

5. Today, the United States flag has 50 stars—one for each state. The first official U.S. flag was approved in 1795. It had $\frac{3}{10}$ as many stars as today's flag. How many stars were on the first official U.S. flag?

   A  5 stars
   B  10 stars
   C  15 stars
   D  35 stars

6. The Statue of Liberty is about 305 feet tall from the ground to the tip of her torch. The statue's pedestal makes up about $\frac{1}{2}$ of its height. About how tall is the pedestal of the Statue of Liberty?

   F  610 feet
   G  152 1/2 feet
   H  150 1/2 feet
   J  102 1/2 feet

7. The Caldwells own a 60-acre farm. They planted $\frac{3}{5}$ of the land with corn. How many acres of corn did they plant?

   A  12 acres
   B  36 acres
   C  20 acres
   D  18 acres

8. Objects on Uranus weigh about $\frac{4}{5}$ of their weight on Earth. If a dog weighs 40 pounds on Earth, how much would it weigh on Uranus?

   F  32 pounds
   G  10 pounds
   H  8 pounds
   J  30 pounds

Name _____ Date _____ Class _____

## LESSON 5-6 Reading Strategies
### Relate Words and Symbols

Repeated addition is a way to represent multiplication of fractions.

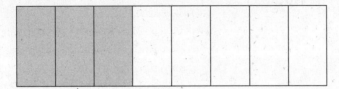

$\frac{1}{8} + \frac{1}{8} + \frac{1}{8} = \frac{3}{8}$     ⟶ Repeated addition

three times one-eighth = three-eighths ⟶ Words

$3 \cdot \frac{1}{8} = \frac{3}{8}$     ⟶ Symbols

**Answer the following questions.**

1. What is $\frac{2}{8} \cdot 2$? _____

2. What is three-eighths times two? _____

3. What is $\frac{1}{8} \cdot 4$? _____

4. Write $\frac{1}{8} + \frac{1}{8} + \frac{1}{8} + \frac{1}{8}$ as a multiplication problem. _____

**Use the rectangle to answer each question.**

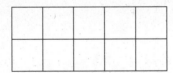

5. What is two-tenths times two? _____

6. What is $\frac{1}{10} \cdot 4$? _____

7. What is four-tenths times two? _____

8. Write $\frac{1}{10} + \frac{1}{10} + \frac{1}{10} + \frac{1}{10}$ as a

multiplication problem in words. _____

Name _____ Date _____ Class _____

## LESSON 5-6 Puzzles, Twisters & Teasers
### Run-Away Computers

Fill in the blanks to complete each statement. Match your answers to the letters to solve the riddle.

1. $\frac{1}{4} \times 16$  _____  T

2. $\frac{1}{6}$ of 18  _____  W

3. $\frac{1}{5}$ of 50  _____  E

4. $\frac{7}{11} \times 22$  _____  R

5. $\frac{4}{5}$ of 15  _____  N

6. $\frac{1}{2} \times 34$  _____  E

7. $\frac{1}{9} \times 45$  _____  V

8. $\frac{3}{4} \times 28$  _____  N

9. $\frac{3}{5} \times 40$  _____  I

10. $\frac{1}{2}$ of 100  _____  T

How do you catch a runaway computer?

With an ___ ___ ___ ___ ___ ___ ___ ___
       24  21  4  10  14  12  17  50

Name _____ Date _____ Class _____

## LESSON 5-7 Practice A
### Multiplying Fractions

**Multiply. Write each answer in simplest form.**

1. $\dfrac{1}{2} \cdot \dfrac{1}{7}$

2. $\dfrac{1}{4} \cdot \dfrac{1}{4}$

3. $\dfrac{1}{5} \cdot \dfrac{1}{3}$

4. $\dfrac{2}{3} \cdot \dfrac{1}{3}$

5. $\dfrac{2}{3} \cdot \dfrac{2}{7}$

6. $\dfrac{1}{4} \cdot \dfrac{1}{5}$

7. $\dfrac{1}{3} \cdot \dfrac{2}{5}$

8. $\dfrac{1}{4} \cdot \dfrac{2}{3}$

9. $\dfrac{1}{3} \cdot \dfrac{1}{3}$

**Evaluate the expression $x \cdot \dfrac{1}{2}$ for each value of $x$. Write the answer in simplest form.**

10. $x = \dfrac{1}{2}$

11. $x = \dfrac{1}{3}$

12. $x = \dfrac{1}{4}$

13. $x = \dfrac{1}{5}$

14. $x = \dfrac{2}{3}$

15. $x = \dfrac{3}{4}$

16. In Mr. Sanders's class, $\dfrac{1}{3}$ of the students are girls. About $\dfrac{1}{4}$ of the girls want to join the chorus. What fraction of all the students in Mr. Sanders's class want to join the chorus?

17. A recipe for trail mix calls for $\dfrac{3}{4}$ pound of peanuts. Luiza only wants to make half of the recipe's servings. How many pounds of peanuts should she use?

Name _____ Date _____ Class _____

## LESSON 5-7 Practice B
### Multiplying Fractions

Multiply. Write each answer in simplest form.

1. $\dfrac{1}{2} \cdot \dfrac{2}{5}$

2. $\dfrac{1}{3} \cdot \dfrac{7}{8}$

3. $\dfrac{2}{3} \cdot \dfrac{4}{6}$

_____  _____  _____

4. $\dfrac{1}{4} \cdot \dfrac{10}{11}$

5. $\dfrac{3}{5} \cdot \dfrac{2}{3}$

6. $\dfrac{8}{9} \cdot \dfrac{3}{4}$

_____  _____  _____

7. $\dfrac{3}{8} \cdot \dfrac{4}{5}$

8. $\dfrac{2}{7} \cdot \dfrac{3}{4}$

9. $\dfrac{1}{6} \cdot \dfrac{2}{3}$

_____  _____  _____

Evaluate the expression $x \cdot \dfrac{1}{5}$ for each value of $x$. Write the answer in simplest form.

10. $x = \dfrac{3}{7}$

11. $x = \dfrac{5}{6}$

12. $x = \dfrac{2}{3}$

_____  _____  _____

13. $x = \dfrac{10}{11}$

14. $x = \dfrac{5}{8}$

15. $x = \dfrac{4}{5}$

_____  _____  _____

16. A cookie recipe calls for $\dfrac{2}{3}$ cup of brown sugar. Sarah is making $\dfrac{1}{4}$ of the recipe. How much brown sugar will she need?

_____

17. Nancy spent $\dfrac{7}{8}$ hour working out at the gym. She spent $\dfrac{5}{7}$ of that time lifting weights. What fraction of an hour did she spend lifting weights?

_____

Name _____ Date _____ Class _____

## LESSON 5-7 Practice C
### Multiplying Fractions

**Multiply. Write each answer in simplest form.**

1. $\dfrac{3}{8} \cdot \dfrac{4}{5}$ _____

2. $\dfrac{5}{8} \cdot \dfrac{3}{9}$ _____

3. $\dfrac{6}{7} \cdot \dfrac{5}{6}$ _____

4. $\dfrac{8}{9} \cdot \dfrac{9}{11}$ _____

5. $\dfrac{5}{12} \cdot \dfrac{6}{7}$ _____

6. $\dfrac{7}{9} \cdot \dfrac{3}{8}$ _____

7. $\dfrac{14}{15} \cdot \dfrac{5}{7}$ _____

8. $\dfrac{7}{8} \cdot \dfrac{2}{9}$ _____

9. $\dfrac{4}{5} \cdot \dfrac{7}{9} \cdot \dfrac{1}{7}$ _____

**Evaluate the expression $x \cdot \dfrac{2}{7}$ for each value of $x$. Write the answer in simplest form.**

10. $x = \dfrac{4}{5}$ _____

11. $x = \dfrac{7}{8}$ _____

12. $x = \dfrac{7}{11}$ _____

13. $x = \dfrac{11}{10}$ _____

14. $x = \dfrac{8}{9}$ _____

15. $x = \dfrac{21}{30}$ _____

**Compare. Write <, >, or =.**

16. $\dfrac{5}{6} \cdot \dfrac{3}{4}\ \square\ \dfrac{7}{8} \cdot \dfrac{4}{5}$

17. $\dfrac{2}{3} \cdot \dfrac{6}{7}\ \square\ \dfrac{9}{10} \cdot \dfrac{1}{3}$

18. $\dfrac{10}{12} \cdot \dfrac{5}{6}\ \square\ \dfrac{5}{9} \cdot \dfrac{1}{4}$

19. $\dfrac{7}{9} \cdot \dfrac{3}{4}\ \square\ \dfrac{7}{6} \cdot \dfrac{1}{2}$

20. $\dfrac{9}{11} \cdot \dfrac{1}{2}\ \square\ \dfrac{1}{6} \cdot \dfrac{3}{11}$

21. $\dfrac{2}{3} \cdot \dfrac{9}{10}\ \square\ \dfrac{4}{5} \cdot \dfrac{7}{8}$

22. Cara bought 1 yard of velvet at the fabric store. She used $\dfrac{5}{9}$ yard to make a purse. Then she used $\dfrac{1}{2}$ of the leftover velvet to make a hair band. How much of the velvet did she use to make the hair band? _____

23. A square-shaped park measures $\dfrac{3}{5}$ mile long on each side. What is the area of the park? _____

Name _____ Date _____ Class _____

## Reteach
### LESSON 5-7 Multiplying Fractions

To multiply fractions, multiply the numerators and multiply the denominators.

When multiplying fractions, you can sometimes divide by the GCF to make the problem simpler.

You can divide by the GCF even if the numerator and denominator of the same fraction have a common factor.

$$\frac{1}{2} \cdot \frac{2}{3}$$

$$\frac{1}{\cancel{2}} \cdot \frac{\cancel{2}}{3}$$

The problem is now $\frac{1}{1} \cdot \frac{1}{3}$.

$$\frac{1 \cdot 1}{1 \cdot 3} = \frac{1}{3}$$

So, $\frac{1}{2} \cdot \frac{2}{3} = \frac{1}{3}$

**Is it possible to simplify before you multiply? If so, what is the GCF?**

1. $\frac{1}{4} \cdot \frac{1}{2}$  2. $\frac{1}{6} \cdot \frac{3}{4}$  3. $\frac{1}{8} \cdot \frac{2}{3}$  4. $\frac{1}{3} \cdot \frac{2}{5}$

_____  _____  _____  _____

**Multiply.**

5. $\frac{1}{6} \cdot \frac{3}{5}$  6. $\frac{1}{4} \cdot \frac{1}{3}$  7. $\frac{7}{8} \cdot \frac{4}{5}$  8. $\frac{1}{6} \cdot \frac{2}{3}$

_____  _____  _____  _____

9. $\frac{1}{5} \cdot \frac{1}{2}$  10. $\frac{3}{5} \cdot \frac{1}{4}$  11. $\frac{3}{7} \cdot \frac{1}{9}$  12. $\frac{3}{4} \cdot \frac{1}{2}$

_____  _____  _____  _____

13. $\frac{1}{3} \cdot \frac{6}{7}$  14. $\frac{1}{4} \cdot \frac{2}{3}$  15. $\frac{3}{4} \cdot \frac{1}{3}$  16. $\frac{1}{4} \cdot \frac{1}{8}$

_____  _____  _____  _____

Name _____ Date _____ Class _____

## LESSON 5-7 Challenge
### Fractions of Flowers

For each flower below, shade the two petals whose fractions have a product equal to the fraction written in the center of that flower.

1.

2.

3.

4.
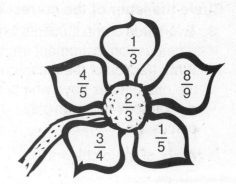

5.

6.

Copyright © by Holt, Rinehart and Winston.
All rights reserved.

Holt Mathematics

Name _____  Date _____  Class _____

## LESSON 5-7 Problem Solving
### Multiplying Fractions

Use the circle graph to answer the questions. Write each answer in simplest form.

1. Of the students playing stringed instruments, $\frac{3}{4}$ play the violin. What fraction of the whole orchestra is violin players?

   _____

2. Of the students playing woodwind instruments, $\frac{1}{2}$ play the clarinet. What fraction of the whole orchestra is clarinet players?

   _____

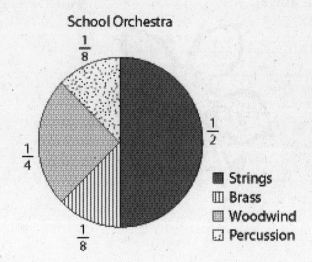

Circle the letter of the correct answer.

3. Two-thirds of the students who play a percussion instrument are boys. What fraction of the musicians in the orchestra is boys who play percussion? girls who play percussion?

   A $\frac{1}{24}$ of the orchestra

   B $\frac{1}{12}$ of the orchestra

   C $\frac{1}{4}$ of the orchestra

   D $\frac{2}{3}$ of the orchestra

4. The brass section is evenly divided into horns, trumpets, trombones, and tubas. What fraction of the whole orchestra do players of each of those brass instruments make up?

   F $\frac{1}{32}$ of the orchestra

   G $\frac{1}{8}$ of the orchestra

   H $\frac{1}{4}$ of the orchestra

   J $\frac{1}{2}$ of the orchestra

5. There are 40 students in the orchestra. How many students play either percussion or brass instruments?

   A 5 students
   B 10 students
   C 8 students
   D 16 students

6. If 2 more violinists join the orchestra, what fraction of all the musicians would play a stringed instrument?

   F $\frac{11}{21}$

   G $\frac{11}{20}$

   H $\frac{1}{20}$

   J $\frac{1}{26}$

Name _____ Date _____ Class _____

## LESSON 5-7 Reading Strategies
### Use Graphic Aids

The circle below is divided into two equal parts. Each part is equal to one-half.

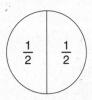

If one-half of the circle is split in half, it looks like this.

$\frac{1}{2}$ of $\frac{1}{2}$ is $\frac{1}{4}$

$\frac{1}{2} \cdot \frac{1}{2} = \frac{1}{4}$

The drawing shows a rectangle divided into thirds.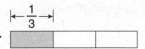

1. If you divide $\frac{1}{3}$ of the rectangle in half, what fractional part will that be? _____

2. One-half of $\frac{1}{3}$ = _____

3. $\frac{1}{2} \cdot \frac{1}{3}$ = _____

To multiply fractions:

$\frac{1}{2} \cdot \frac{1}{4}$

$\frac{2 \cdot 1}{3 \cdot 4} = \frac{2}{12}$ ← Multiply numerators.
← Multiply denominators.

$\frac{2}{12} = \frac{1}{6}$ ← Answer in simplest form

Use the problem $\frac{2}{5} \cdot \frac{3}{4}$ to answer the following questions.

4. When you multiply the numerators, the product is _____.

5. When you multiply the denominators, the product is _____.

6. $\frac{2}{5} \cdot \frac{3}{4}$ = _____

Name _____ Date _____ Class _____

## LESSON 5-7 Puzzles, Twisters & Teasers
### Itchy Multiplication

Solve each problem and find the answer in the box. Place the letter corresponding to the answer in the blanks to answer the riddle.

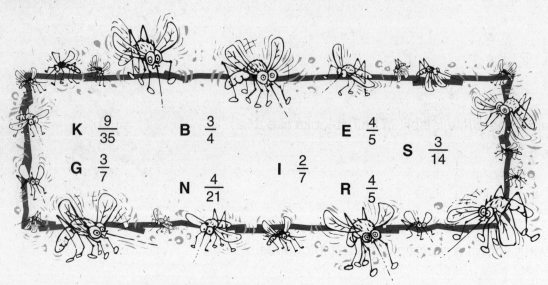

K $\frac{9}{35}$   B $\frac{3}{4}$   E $\frac{4}{5}$   S $\frac{3}{14}$
G $\frac{3}{7}$   N $\frac{4}{21}$   I $\frac{2}{7}$   R $\frac{4}{5}$

Find the value of each expression if $n = \frac{3}{7}$.

1. The value of $\frac{4}{9}n$ _____

2. The value of $\frac{2}{3}n$ _____

3. The value of $\frac{3}{5}n$ _____

4. The value of $\frac{1}{2}n$ _____

What is a mosquito's favorite sport? ___ ___ ___ ___ DIVING
                                                                   4   3   2   1

Name _____ Date _____ Class _____

## LESSON 5-8 Practice A
### Multiplying Mixed Numbers

**Multiply. Write each answer in simplest form.**

1. $1\frac{1}{2} \cdot 1\frac{1}{3}$

   $\frac{1}{2} \cdot \frac{\phantom{3}}{3}$

   _____

2. $1\frac{1}{5} \cdot \frac{4}{5}$

   $\frac{\phantom{5}}{5} \cdot \frac{4}{5}$

   _____

3. $1\frac{1}{4} \cdot \frac{2}{3}$

   $\frac{\phantom{4}}{4} \cdot \frac{2}{3}$

   _____

4. $1\frac{1}{8} \cdot \frac{2}{5}$

   $\frac{\phantom{8}}{8} \cdot \frac{2}{5}$

   _____

5. $\frac{2}{5} \cdot 1\frac{1}{2}$

   $\frac{2}{5} \cdot \frac{\phantom{2}}{2}$

   _____

6. $1\frac{3}{5} \cdot \frac{1}{3}$

   $\frac{\phantom{5}}{5} \cdot \frac{1}{3}$

   _____

7. $\frac{2}{7} \cdot 1\frac{1}{4}$

   _____

8. $\frac{2}{3} \cdot 1\frac{1}{10}$

   _____

9. $\frac{1}{8} \cdot 1\frac{1}{2}$

   _____

**Find each product. Write the answer in simplest form.**

10. $\frac{4}{5} \cdot 1\frac{1}{6}$

    _____

11. $\frac{3}{5} \cdot 1\frac{1}{4}$

    _____

12. $1\frac{3}{4} \cdot \frac{1}{3}$

    _____

13. $2 \cdot 1\frac{1}{2}$

    _____

14. $4 \cdot 2\frac{1}{4}$

    _____

15. $5 \cdot 1\frac{1}{5}$

    _____

16. Lin Li makes two and a half dollars per hour baby-sitting her little brother. How much money will she make if she baby-sits for 5 hours?

_____

17. Andrea is baking 2 batches of cookies. The recipe calls for $4\frac{1}{2}$ cups of flour for each batch. How many cups of flour will she use?

_____

Name _____ Date _____ Class _____

## Practice B
**LESSON 5-8** *Multiplying Mixed Numbers*

**Multiply. Write each answer in simplest form.**

1. $1\frac{2}{3} \cdot \frac{4}{5}$

2. $1\frac{7}{8} \cdot \frac{4}{5}$

3. $2\frac{3}{4} \cdot \frac{1}{5}$

_____   _____   _____

4. $2\frac{1}{6} \cdot \frac{2}{3}$

5. $2\frac{2}{5} \cdot \frac{3}{8}$

6. $1\frac{3}{4} \cdot \frac{5}{6}$

_____   _____   _____

7. $1\frac{1}{6} \cdot \frac{3}{5}$

8. $\frac{2}{9} \cdot 2\frac{1}{7}$

9. $2\frac{3}{11} \cdot \frac{7}{10}$

_____   _____   _____

**Find each product. Write the answer in simplest form.**

10. $\frac{6}{7} \cdot 1\frac{1}{4}$

11. $\frac{5}{8} \cdot 1\frac{3}{5}$

12. $2\frac{4}{9} \cdot \frac{1}{6}$

_____   _____   _____

13. $1\frac{3}{10} \cdot 1\frac{1}{3}$

14. $2\frac{1}{2} \cdot 2\frac{1}{2}$

15. $1\frac{2}{3} \cdot 3\frac{1}{2}$

_____   _____   _____

16. Dominick lives $1\frac{3}{4}$ miles from his school. If his mother drives him half the way, how far will Dominick have to walk to get to school?

_____

17. Katoni bought $2\frac{1}{2}$ dozen donuts to bring to the office. Since there are 12 donuts in a dozen, how many donuts did Katoni buy?

_____

Name _____ Date _____ Class _____

## LESSON 5-8 Practice C
### Multiplying Mixed Numbers

**Multiply. Write each answer in simplest form.**

1. $\frac{5}{9} \cdot 2\frac{2}{7}$

2. $1\frac{11}{12} \cdot \frac{6}{7}$

3. $2\frac{4}{9} \cdot \frac{7}{8}$

4. $3\frac{2}{3} \cdot \frac{3}{5}$

5. $\frac{13}{14} \cdot 1\frac{3}{4}$

6. $2\frac{3}{10} \cdot \frac{5}{6}$

7. $1\frac{7}{8} \cdot \frac{3}{5}$

8. $3\frac{2}{7} \cdot \frac{3}{10}$

9. $4\frac{2}{3} \cdot \frac{8}{9}$

**Find each product. Write the answer in simplest form.**

10. $\frac{10}{11} \cdot 3\frac{3}{7} \cdot 2$

11. $2\frac{4}{7} \cdot \frac{4}{5} \cdot 1\frac{1}{2}$

12. $\frac{9}{12} \cdot 2\frac{3}{5} \cdot 3\frac{1}{4}$

13. $6\frac{1}{5} \cdot 10 \cdot 3\frac{4}{5}$

14. $1\frac{7}{9} \cdot \frac{2}{5} \cdot 5\frac{1}{10}$

15. $2\frac{6}{7} \cdot 1\frac{8}{9} \cdot \frac{7}{8}$

**Evaluate each expression.**

16. $\frac{3}{4} \cdot c$ for $c = 4\frac{4}{5}$

17. $1\frac{3}{10} \cdot x$ for $x = 2\frac{2}{3}$

18. $\frac{2}{9} \cdot h$ for $h = 3\frac{5}{6}$

19. $\frac{3}{4} \cdot q$ for $q = 2\frac{7}{8}$

20. A train travels at $110\frac{3}{10}$ miles per hour. At this rate, how far will the train travel in $2\frac{1}{2}$ hours?

21. A sandbox is $1\frac{1}{3}$ feet tall, $1\frac{5}{8}$ feet wide, and $4\frac{1}{2}$ feet long. How many cubic feet of sand is needed to fill the box?
(Volume = length • width • height)

# Reteach
## 5-8 Multiplying Mixed Numbers

To find $\frac{1}{3}$ of $2\frac{1}{2}$, first change $2\frac{1}{2}$ to an improper fraction.

$2\frac{1}{2} = \frac{5}{2}$

Then multiply as you would with two proper fractions.

Check to see if you can divide by the GCF to make the problem simpler. Then multiply the numerators and multiply the denominators.

The problem is now $\frac{1}{3} \cdot \frac{5}{2}$.

$\frac{1 \cdot 5}{3 \cdot 2} = \frac{5}{6}$

So, $\frac{1}{3} \cdot 2\frac{1}{2}$ is $\frac{5}{6}$.

**Rewrite each mixed number as an improper fraction. Is it possible to simplify before you multiply? If so, what is the GCF?**

1. $\frac{1}{4} \cdot 1\frac{1}{3}$
   
   $= \frac{1}{4} \cdot \underline{\phantom{xx}}$

2. $\frac{1}{6} \cdot 2\frac{1}{2}$
   
   $= \frac{1}{6} \cdot \underline{\phantom{xx}}$

3. $\frac{1}{8} \cdot 1\frac{1}{2}$
   
   $= \frac{1}{8} \cdot \underline{\phantom{xx}}$

4. $\frac{1}{3} \cdot 1\frac{2}{5}$
   
   $= \frac{1}{3} \cdot \underline{\phantom{xx}}$

5. $1\frac{1}{3} \cdot 1\frac{2}{3}$
   
   $\frac{\phantom{x}}{3} \cdot \frac{\phantom{x}}{3}$

6. $1\frac{1}{2} \cdot 1\frac{1}{3}$
   
   $\frac{\phantom{x}}{2} \cdot \frac{\phantom{x}}{3}$

7. $1\frac{3}{4} \cdot 2\frac{1}{2}$
   
   $\frac{\phantom{x}}{4} \cdot \frac{\phantom{x}}{2}$

8. $1\frac{1}{6} \cdot 2\frac{2}{3}$
   
   $\frac{\phantom{x}}{6} \cdot \frac{\phantom{x}}{3}$

9. $3\frac{1}{3} \cdot \frac{2}{5}$

10. $2\frac{1}{2} \cdot \frac{1}{5}$

11. $1\frac{3}{4} \cdot 2\frac{1}{2}$

12. $3\frac{1}{3} \cdot 1\frac{1}{5}$

Name _____ Date _____ Class _____

## LESSON 5-8 Challenge

### And They're Off!

Like many sports, horse racing uses a special system of measurement. Horse races are measured in units called *furlongs*. One furlong equals $\frac{1}{8}$ mile. The races described below have different furlong lengths, but they all offer the same prize money to their winners—$1,000,000!

**Write the length in miles of each of these horse races in simplest form.**

1. Santa Anita Derby, California

   Race Length: 9 furlongs          Length in Miles: _____

2. Kentucky Derby, Kentucky

   Race Length: 10 furlongs         Length in Miles: _____

3. Preakness Stakes, Maryland

   Race Length: $9\frac{1}{2}$ furlongs        Length in Miles: _____

4. Belmont Stakes, New York

   Race Length: 12 furlongs         Length in Miles: _____

5. Breeders' Cup Juvenile, New York

   Race Length: $8\frac{1}{2}$ furlongs        Length in Miles: _____

## Problem Solving
### 5-8 Multiplying Mixed Numbers

Use the recipe to answer the questions.

| CHOCOLATE CHIP COOKIES |
| --- |
| Servings: 1 batch |
| $1\frac{2}{3}$ cups flour |
| $\frac{3}{4}$ teaspoon baking soda |
| $\frac{1}{2}$ cup white sugar |
| $2\frac{1}{3}$ cups semisweet chocolate chips |
| $\frac{1}{2}$ cup brown sugar |
| $\frac{3}{4}$ cup butter |
| 1 egg |
| $1\frac{1}{4}$ teaspoons vanilla |

1. If you want to make $2\frac{1}{2}$ batches, how much flour would you need?

    _____

2. If you want to make only $1\frac{1}{2}$ batches, how much chocolate chips would you need?

    _____

3. You want to bake $3\frac{1}{4}$ batches. How much vanilla do you need in all?

    _____

Choose the letter for the best answer.

4. If you make $1\frac{1}{4}$ batches, how much baking soda would you need?

    A $\frac{3}{16}$ teaspoon  C $\frac{3}{5}$ teaspoon
    B $\frac{5}{16}$ teaspoon  D $\frac{15}{16}$ teaspoon

5. How many cups of white sugar do you need to make $3\frac{1}{2}$ batches of cookies?

    F $3\frac{1}{2}$ cups  H $1\frac{1}{2}$ cups
    G $1\frac{3}{4}$ cups  J $1\frac{1}{4}$ cups

6. Dan used $2\frac{1}{4}$ cups of butter to make chocolate chip cookies using the above recipe. How many batches of cookies did he make?

    A 3 batches  C 5 batches
    B 4 batches  D 6 batches

7. One bag of chocolate chips holds 2 cups. If you buy five bags, how many cups of chips will you have left over after baking $2\frac{1}{2}$ batches of cookies?

    F $4\frac{1}{6}$ cups  H $2\frac{1}{3}$ cups
    G $5\frac{5}{6}$ cups  J $\frac{1}{3}$ cup

Name _____ Date _____ Class _____

# Reading Strategies
## LESSON 5-8 Use a Flow Chart

**Mixed Number**

Whole number → $2\frac{1}{2}$ ← Fraction

**Improper Fraction**

$$\frac{5}{2}$$

You can change mixed numbers to improper fractions.

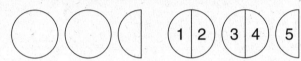

$2\frac{1}{2}$ = 5 halves or $\frac{5}{2}$ ← improper fraction

1. What is the mixed number in the above example? _____

2. What is the improper fraction? _____

3. How many halves are in $2\frac{1}{2}$? _____

**Use the flowchart below to help you change a mixed number to an improper fraction.**

| Multiply the denominator by the whole number. | → | Add the numerator. | → | The denominator stays the same. |

**Change $3\frac{2}{5}$ to an improper fraction.**

4. What is the first step?

_____

5. What is the next step?

_____

6. The improper fraction is _____.

Copyright © by Holt, Rinehart and Winston.
All rights reserved.

Holt Mathematics

Name _____ Date _____ Class _____

**LESSON 5-8**

# Puzzles, Twisters & Teasers
## All Mixed Up!

Rami was carrying a set of cards, but he tripped. The cards fell on the floor and are all mixed up. Help Rami put them in order by solving each problem.

Once you have solved the problems, place the cards in order from least to greatest. When in order, the letters will spell out a message!

| B | O | J | O |
|---|---|---|---|
| $6 \cdot 2\frac{2}{3}$ | $3\frac{1}{4} \cdot 3\frac{2}{5}$ | $2\frac{3}{4} \cdot 3\frac{2}{3}$ | $1\frac{1}{2} \cdot 4\frac{5}{6}$ |

| D | O | G |
|---|---|---|
| $5\frac{1}{2} \cdot 1\frac{2}{5}$ | $\frac{4}{5} \cdot 3\frac{5}{6}$ | $\frac{5}{7} \cdot \frac{1}{8}$ |

The message is… _____

Name _____ Date _____ Class _____

## LESSON 5-9 Practice A
### Dividing Fractions and Mixed Numbers

**Find the reciprocal.**

1. $\frac{1}{2}$  2. $\frac{2}{3}$  3. $\frac{1}{5}$

_____  _____  _____

4. $\frac{1}{3}$  5. $\frac{3}{5}$  6. $1\frac{1}{4}$

_____  _____  _____

7. $\frac{2}{5}$  8. $\frac{3}{7}$  9. $1\frac{1}{2}$

_____  _____  _____

**Divide. Write each answer in simplest form.**

10. $\frac{2}{3} \div 2$  11. $\frac{1}{2} \div \frac{3}{4}$  12. $\frac{5}{6} \div \frac{1}{4}$

$\frac{2}{3} \cdot$ _____  $\frac{1}{2} \cdot$ _____  $\frac{5}{6} \cdot$ _____

_____  _____  _____

13. $\frac{3}{5} \div \frac{1}{5}$  14. $\frac{7}{9} \div 3$  15. $1\frac{1}{2} \div \frac{1}{2}$

$\frac{3}{5} \cdot$ _____  $\frac{7}{9} \cdot$ _____  $1\frac{1}{2} \cdot$ _____

_____  _____  _____

16. Stella has 6 pounds of chocolate. She will use $\frac{2}{3}$ pound of the chocolate to make one cake. How many cakes can she make? _____

17. Todd has $\frac{8}{9}$ pound of clay. He will use $\frac{1}{3}$ pound to make each action figure. How many action figures can he make? _____

18. Dylan gives his two guinea pigs a total of $\frac{3}{4}$ cup of food every day. If each guinea pig gets the same amount of food, how much do they each get each day? _____

## Practice B
### 5-9 Dividing Fractions and Mixed Numbers

**Find the reciprocal.**

1. $\dfrac{5}{7}$ _____

2. $\dfrac{9}{8}$ _____

3. $\dfrac{3}{5}$ _____

4. $\dfrac{1}{10}$ _____

5. $\dfrac{4}{9}$ _____

6. $\dfrac{13}{14}$ _____

7. $1\dfrac{1}{3}$ _____

8. $2\dfrac{4}{5}$ _____

9. $3\dfrac{1}{6}$ _____

**Divide. Write each answer in simplest form.**

10. $\dfrac{5}{6} \div 5$ _____

11. $2\dfrac{3}{4} \div 1\dfrac{4}{7}$ _____

12. $\dfrac{7}{8} \div \dfrac{2}{3}$ _____

13. $3\dfrac{1}{4} \div 2\dfrac{3}{4}$ _____

14. $\dfrac{9}{10} \div 3$ _____

15. $\dfrac{3}{4} \div 9$ _____

16. $2\dfrac{6}{9} \div \dfrac{6}{7}$ _____

17. $\dfrac{5}{6} \div 2\dfrac{3}{10}$ _____

18. $2\dfrac{1}{8} \div 3\dfrac{1}{4}$ _____

19. The rope in the school gymnasium is $10\dfrac{1}{2}$ feet long. To make it easier to climb, the gym teacher tied a knot in the rope every $\dfrac{3}{4}$ foot. How many knots are in the rope? _____

20. Mr. Fulton bought $12\dfrac{1}{2}$ pounds of ground beef for the cookout. He plans on using $\dfrac{1}{4}$ pound of beef for each hamburger. How many hamburgers can he make? _____

21. Mrs. Marks has $9\dfrac{1}{4}$ ounces of fertilizer for her plants. She plans on using $\dfrac{3}{4}$ ounce of fertilizer for each plant. How many plants can she fertilizer? _____

Name _____ Date _____ Class _____

## LESSON 5-9 Practice C
### Dividing Fractions and Mixed Numbers

**Find the reciprocal.**

1. $10\tfrac{1}{2}$ 

2. $6\tfrac{3}{7}$

3. $2\tfrac{8}{9}$

4. $15\tfrac{1}{4}$

5. $9\tfrac{2}{3}$

6. $7\tfrac{5}{8}$

**Divide. Write each answer in simplest form.**

7. $\tfrac{8}{10} \div 1\tfrac{5}{6}$

8. $\tfrac{8}{9} \div \tfrac{6}{7}$

9. $3\tfrac{3}{5} \div 2\tfrac{1}{4}$

10. $4\tfrac{1}{2} \div 2\tfrac{3}{8}$

11. $5\tfrac{5}{6} \div 3\tfrac{1}{6}$

12. $\tfrac{11}{12} \div 2\tfrac{5}{8}$

13. $1\tfrac{9}{13} \div \tfrac{3}{8}$

14. $6\tfrac{4}{5} \div 3\tfrac{2}{9}$

15. $8\tfrac{2}{11} \div 2\tfrac{4}{7}$

16. $9\tfrac{6}{13} \div 10$

17. $12\tfrac{1}{3} \div 5\tfrac{4}{5}$

18. $9\tfrac{2}{3} \div 6\tfrac{8}{9}$

19. The area of the public swimming pool is $510\tfrac{7}{8}$ square feet. The pool is $30\tfrac{1}{2}$ feet long. What is the width of the pool?

20. At the bank, Pamela exchanged all of her quarters for 16 five-dollar bills. How many quarters did Pamela exchange?

21. Barbara has $16\tfrac{1}{5}$ yards of fabric. She will use $5\tfrac{2}{5}$ yards to make each costume. How many costumes can Barbara make?

Holt Mathematics

# LESSON 5-9 Reteach
## Dividing Fractions and Mixed Numbers

Two numbers are reciprocals if their product is 1. $\frac{2}{3}$ and $\frac{3}{2}$ are reciprocals because $\frac{2}{3} \cdot \frac{3}{2} = \frac{6}{6} = 1$.

Dividing by a fraction is the same as multiplying by its reciprocal.

$\frac{1}{4} \div 2 = \frac{1}{8}$       $\frac{1}{4} \cdot \frac{1}{2} = \frac{1}{8}$

So, you can use reciprocals to divide by fractions.

To find $\frac{2}{3} \div 4$, first rewrite the expression as a multiplication expression using the reciprocal of the divisor, 4.

$\frac{2}{3} \cdot \frac{1}{4}$

Then use canceling to find the product in simplest form.

$\frac{2}{3} \div 4 = \frac{2}{3} \cdot \frac{1}{4} = \frac{1}{3} \cdot \frac{1}{2} = \frac{1}{6}$

To find $3\frac{1}{4} \div 1\frac{1}{2}$, first rewrite the expression using improper fractions.

$\frac{13}{4} \div \frac{3}{2}$

Next, write the expression as a multiplication expression.

$\frac{13}{4} \cdot \frac{2}{3}$

$3\frac{1}{4} \div 1\frac{1}{2} = \frac{13}{4} \div \frac{3}{2} = \frac{13}{4} \cdot \frac{2}{3} = \frac{13}{2} \cdot \frac{1}{3} = \frac{13}{6} = 2\frac{1}{6}$

**Divide. Write each answer in simplest form.**

1. $\frac{1}{4} \div 3$

   $\frac{1}{4} \div \frac{\phantom{0}}{1}$

   $\phantom{0} \cdot \phantom{0}$

   _____

2. $1\frac{1}{2} \div 1\frac{1}{4}$

   $\frac{3}{2} \div \frac{\phantom{0}}{4}$

   $\phantom{0} \cdot \phantom{0}$

   _____

3. $\frac{3}{8} \div 2$

   $\frac{3}{8} \div \frac{\phantom{0}}{1}$

   $\phantom{0} \cdot \phantom{0}$

   _____

4. $2\frac{1}{3} \div 1\frac{3}{4}$

   $\frac{\phantom{0}}{3} \div \frac{\phantom{0}}{4}$

   $\phantom{0} \cdot \phantom{0}$

   _____

5. $\frac{1}{5} \div 2$

6. $1\frac{1}{6} \div 2\frac{2}{3}$

7. $\frac{1}{8} \div 4$

8. $3\frac{1}{8} \div \frac{1}{2}$

Name _____ Date _____ Class _____

## LESSON 5-9 Challenge
### Inching Across the U.S.A.

You can use a map and a ruler to find the distance between places. On the map below, for example, you measure that 2 inches separate Kansas City, Missouri, and Richmond, Virginia. The map scale shows that $\frac{1}{4}$ inch on the map equals 140 miles. So the distance between Kansas City and Richmond is 1,120 miles.

**Calculations:**

$2 \div \frac{1}{4} = 8$

$8 \times 140 = 1{,}120$

**Use the map and a ruler to find the distance in miles between each pair of cities.**

1. Miami, Florida, and New Orleans, Louisiana _____
2. Denver, Colorado, and Los Angeles, California _____
3. Seattle, Washington, and Minneapolis, Minnesota _____
4. Washington, D.C., and Atlanta, Georgia _____
5. Oklahoma City, Oklahoma, and Pittsburgh, Pennsylvania _____
6. San Francisco, California, and Boston, Massachusetts _____

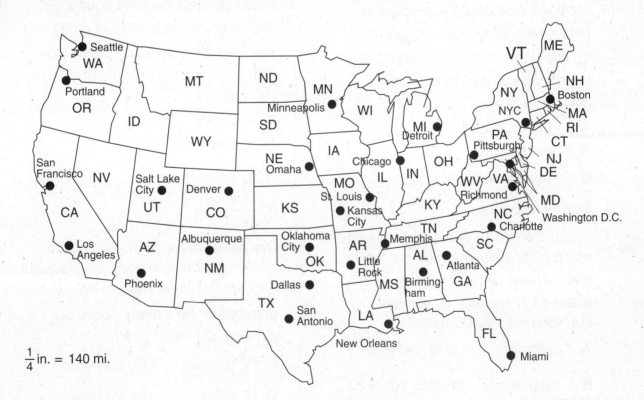

$\frac{1}{4}$ in. = 140 mi.

# Problem Solving
## 5-9 Dividing Fractions and Mixed Numbers

**Write the correct answer in simplest form.**

1. Horses are measured in units called *hands*. One inch equals $\frac{1}{4}$ hand. The average Clydesdale horse is $17\frac{1}{5}$ hands high. What is the horse's height in inches? in feet?

2. Cloth manufacturers use a unit of measurement called a *finger*. One finger is equal to $4\frac{1}{2}$ inches. If 25 inches are cut off a bolt of cloth, how many fingers of cloth were cut?

3. People in England measure weights in units called *stones*. One pound equals $\frac{1}{14}$ of a stone. If a cat weighs $\frac{3}{4}$ stone, how many pounds does it weigh?

4. The hiking trail is $\frac{9}{10}$ mile long. There are 6 markers evenly posted along the trail to direct hikers. How far apart are the markers placed?

**Choose the letter for the best answer.**

5. A cake recipe calls for $1\frac{1}{2}$ cups of butter. One tablespoon equals $\frac{1}{16}$ cup. How many tablespoons of butter do you need to make the cake?
   A  24 tablespoons
   B  8 tablespoons
   C  $\frac{3}{32}$ tablespoon
   D  9 tablespoons

6. Printed letters are measured in units called *points*. One point equals $\frac{1}{72}$ inch. If you want the title of a paper you are typing on a computer to be $\frac{1}{2}$ inch tall, what type point size should you use?
   F  144 point
   G  36 point
   H  $\frac{1}{36}$ point
   J  $\frac{1}{144}$ point

7. Phyllis bought 14 yards of material to make chair cushions. She cut the material into pieces $1\frac{3}{4}$ yards long to make each cushion. How many cushions did Phyllis make?
   A  4 cushions      C  8 cushions
   B  6 cushions      D  $24\frac{1}{2}$ cushions

8. Dry goods are sold in units called *pecks* and *bushels*. One peck equals $\frac{1}{4}$ bushel. If Peter picks $5\frac{1}{2}$ bushels of peppers, how many pecks of peppers did Peter pick?
   F  $1\frac{3}{8}$ pecks    H  20 pecks
   G  11 pecks         J  22 pecks

Name _____ Date _____ Class _____

## LESSON 5-9 Reading Strategies
### Using Models

Fraction bars help you picture dividing by fractions.

[ ]   [ ]   [ $\frac{1}{2}$ ]

[ $\frac{1}{4}$ ]

In the problem $2\frac{1}{2} \div \frac{1}{4}$, think: How many one-fourths are there in $2\frac{1}{2}$?

| $\frac{1}{4}$ | $\frac{1}{4}$ | $\frac{1}{4}$ | $\frac{1}{4}$ |   | $\frac{1}{4}$ | $\frac{1}{4}$ | $\frac{1}{4}$ | $\frac{1}{4}$ |   | $\frac{1}{4}$ | $\frac{1}{4}$ |

**Use the picture to answer each question.**

1. Count the number of $\frac{1}{4}$'s in the fraction bars above. How many are there? _____

2. $2\frac{1}{2} \div \frac{1}{4} =$ _____

In the problem $2\frac{1}{2} \times 4$, think $2\frac{1}{2}$ four times.

[ ]   [ ]
[ ]   [ ]
[ $\frac{1}{2}$ ]   [ $\frac{1}{2}$ ]
[ ]   [ ]
[ ]   [ ]
[ $\frac{1}{2}$ ]   [ $\frac{1}{2}$ ]

**Use the picture to answer each question.**

3. How many whole fraction bars are there? _____

4. How many one-half fraction bars are there? _____

5. When you add the whole bars and half bars together you get _____ whole bars.

6. Compare the multiplication and division examples. What do you notice about the answer you get when you divide by $\frac{1}{4}$ or multiply by 4?

_____

Name _____ Date _____ Class _____

## LESSON 5-9 Puzzles, Twisters & Teasers
### Divide and Conquer!

You've heard there is a Pot of Gold to be found. Begin at "S"(start). Decide whether the first statement is true or false. Circle your answer, and move as directed. Go to problem 2. Decide whether the statement is true or false and move as directed. Can you conquer the maze and make it to the Pot of Gold?

1. The reciprocal of $\frac{5}{7}$ is $\frac{7}{5}$.

   True: 7 steps right and 6 steps down.

   False: 6 steps right and 7 steps down.

2. $\frac{2}{3} \div \frac{1}{3} = \frac{2}{9}$

   True: 3 steps right and 7 steps up.

   False: 7 steps right and 3 steps up.

3. The reciprocal of $4\frac{3}{5}$ is $4\frac{5}{3}$.

   True: 6 steps right and 3 steps down.

   False: 3 steps right and 6 steps down.

4. $3\frac{1}{2} + 2\frac{1}{6} = 5\frac{2}{3}$

   True: 2 steps diagonally up and to the right.

   False: 2 steps diagonally down and to the right.

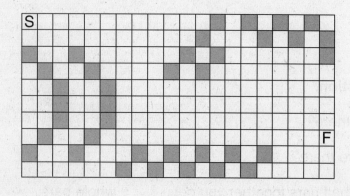

Copyright © by Holt, Rinehart and Winston.
All rights reserved.

74

Holt Mathematics

Name _____ Date _____ Class _____

## LESSON 5-10 Practice A
### Solving Fraction Equations: Multiplication and Division

Solve each equation. Write the answer in simplest form.

1. $\frac{1}{2}x = 2$

2. $2t = \frac{2}{3}$

3. $\frac{1}{3}a = 3$

_____  _____  _____

4. $\frac{r}{2} = 4$

5. $\frac{b}{3} = 6$

6. $2y = \frac{1}{5}$

_____  _____  _____

7. $\frac{1}{4}d = 2$

8. $\frac{b}{5} = 6$

9. $\frac{q}{10} = \frac{1}{5}$

_____  _____  _____

10. $\frac{1}{3}s = 4$

11. $\frac{h}{2} = 2$

12. $\frac{1}{4}c = 1$

_____  _____  _____

**Circle the correct answer.**

13. Tate earned $9 for working $\frac{3}{4}$ of an hour. Which equation can be used to find Tate's hourly rate?

 A $9h = \frac{3}{4}$

 B $9 + \frac{3}{4} = h$

 C $\frac{3}{4}h = 9$

 D $9 - \frac{3}{4} = h$

14. Which operation should you use to solve the equation $5x = 2$?

 F addition

 G subtraction

 H multiplication

 J division

15. A number $n$ is divided by 2, and the quotient is $\frac{1}{3}$. Write an equation to model this problem.

_____

16. A number $n$ is multiplied by $\frac{1}{4}$, and the product is 5. Write and solve an equation to model this problem.

_____

# Practice B
## 5-10 Solving Fraction Equations: Multiplication and Division

Solve each equation. Write the answer in simplest form. Check your answers.

1. $\frac{1}{4}x = 6$

2. $2t = \frac{4}{7}$

3. $\frac{3}{5}a = 3$

4. $\frac{r}{6} = 8$

5. $\frac{2b}{9} = 4$

6. $3y = \frac{4}{5}$

7. $\frac{2}{3}d = 5$

8. $2f = \frac{1}{6}$

9. $4q = \frac{2}{9}$

10. $\frac{1}{2}s = 2$

11. $\frac{h}{7} = 5$

12. $\frac{1}{4}c = 9$

13. $5g = \frac{5}{6}$

14. $3k = \frac{1}{9}$

15. $\frac{3x}{5} = 6$

16. It takes 3 buckets of water to fill $\frac{1}{3}$ of a fish tank. How many buckets are needed to fill the whole tank? _____

17. Jenna got 12, or $\frac{3}{5}$, of her answers on the test right. How many questions were on the test? _____

18. It takes Charles 2 minutes to run $\frac{1}{4}$ of a mile. How long will it take Charles to run a mile? _____

Name _____ Date _____ Class _____

## LESSON 5-10 Practice C
### Solving Fraction Equations: Multiplication and Division

Solve each equation. Write the answer in simplest form. Check your answers.

1. $\frac{2}{3}x = 10$

2. $5t = \frac{10}{15}$

3. $\frac{6}{7}a = 9$

4. $\frac{r}{11} = 12$

5. $\frac{6b}{9} = 15$

6. $7y = \frac{7}{8}$

7. $\frac{4}{5}d = 15$

8. $4f = \frac{1}{9}$

9. $7q = \frac{3}{5}$

10. $\frac{7}{8}s = 14$

11. $\frac{h}{12} = 6$

12. $\frac{3}{10}c = \frac{2}{3}$

13. $\frac{5g}{6} = \frac{7}{12}$

14. $\frac{3k}{9} = \frac{5}{6}$

15. $5\frac{1}{2}n = 3$

16. Anya worked $8\frac{1}{4}$ hours on Saturday and $6\frac{1}{4}$ hours on Sunday. She earned a total of $137.75 for both days combined. How much does Anya make per hour? _____

17. Ernest rode his bike $6\frac{1}{4}$ miles on Saturday and $8\frac{1}{2}$ miles on Sunday. He rode for a total of $88\frac{1}{2}$ minutes for both days combined. How long does it take him to ride a mile on his bike? _____

# Reteach
## 5-10 Solving Fraction Equations: Multiplication and Division

You can write related facts using multiplication and division.

$3 \cdot 4 = 12 \qquad 4 = 12 \div 3$

You can use related facts to solve equations.

**A.** $\frac{2}{3} \cdot x = 12$

Think: $12 \div \frac{2}{3} = x$

$x = 12 \cdot \frac{3}{2}$

$x = \frac{12}{1} \cdot \frac{3}{2}$

$x = \frac{36}{2}$

$x = 18$

**Check:** $\frac{2}{3} \cdot x = 12$

$\frac{2}{3} \cdot 18 \stackrel{?}{=} 12$ Substitute

$\frac{2}{3} \cdot \frac{18}{1} \stackrel{?}{=} 12$

$\frac{36}{3} \stackrel{?}{=} 12$

$12 = 12$ ✓

**B.** $\frac{2x}{5} = 3$

$\frac{2}{5} \cdot x = 3$

Think: $3 \div \frac{2}{5} = x$

$x = 3 \cdot \frac{5}{2}$

$x = \frac{3}{1} \cdot \frac{5}{2}$

$x = \frac{15}{2}$

$x = 7\frac{1}{2}$

**Check:** $\frac{2x}{5} = 3$

$\frac{2}{5} \cdot x \stackrel{?}{=} 3$

$\frac{2}{5} \cdot \frac{15}{2} \stackrel{?}{=} 3$ Substitute

$\frac{30}{10} \stackrel{?}{=} 3$

$3 = 3$ ✓

Use related facts to solve each equation. Then check each answer.

1. $\frac{1}{4} \cdot x = 3$

2. $\frac{3x}{4} = 2$

3. $\frac{3}{5} \cdot x = \frac{2}{3}$

_____

_____

_____

4. $\frac{1}{3} \cdot x = 6$

5. $\frac{2x}{5} = 1$

6. $\frac{1}{8} \cdot x = 3$

_____

_____

_____

Name _____ Date _____ Class _____

## Challenge
### LESSON 5-10 *Crawly Creature Equations*

A millipede called the *Illacme plenipes* holds the record for the creature with the most legs—750! However, most millipedes have only 30 legs. Shown below are some other many-legged creatures.

Let  = the number of legs most millipedes have. Use this information to solve the equations and find how many legs each other crawly creature has.

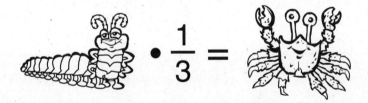

$\dfrac{8}{15} \cdot$ 🐛 $=$ 🐛

$\dfrac{3}{5} \cdot$ 🐛 $\cdot \dfrac{5}{6} =$ 🕷

$\dfrac{3}{4} \cdot$ 🕷 $=$ 🪰

🐛 $\cdot \dfrac{1}{3} =$ 🦀

**Caterpillars**   **Spiders**   **Insects**   **Crabs**

_____   _____   _____   _____

## Problem Solving
### 5-10 Solving Fraction Equations: Multiplication and Division

**Solve.**

1. The number of T-shirts is multiplied by $\frac{1}{2}$ and the product is 18. Write and solve an equation for the number of T-shirts, where $t$ represents the number of T-shirts.

   _____

2. The number of students is divided by 18 and the quotient is $\frac{1}{6}$. Write and solve an equation for the number of students, where $s$ represents the number of students.

   _____

3. The number of players is multiplied by $2\frac{1}{2}$ and the product is 25. Write and solve an equation for the number of players, where $p$ represents the number of players.

   _____

4. The number of chairs is divided by $\frac{1}{4}$ and the quotient is 12. Write and solve an equation for the number of chairs, where $c$ represents the number of chairs.

   _____

**Circle the letter of the correct answer.**

5. Paco bought 10 feet of rope. He cut it into several $\frac{5}{6}$-foot pieces. Which equation can you use to find how many pieces of rope Paco cut?

   A  $\frac{5}{6} \div 10 = x$

   B  $\frac{5}{6} \div x = 10$

   C  $10 \div x = \frac{5}{6}$

   D  $10x = \frac{5}{6}$

6. Each square on the graph paper has an area of $\frac{4}{9}$ square inch. What is the length and width of each square?

   F  $\frac{1}{9}$ inch

   G  $\frac{2}{3}$ inch

   H  $\frac{2}{9}$ inch

   J  $\frac{1}{3}$ inch

7. Which operation should you use to solve the equation $6x = \frac{3}{8}$?

   A  addition
   B  subtraction
   C  multiplication
   D  division

8. A fraction divided by $\frac{2}{3}$ is equal to $1\frac{1}{4}$. What is that fraction?

   F  $\frac{1}{3}$

   G  $\frac{5}{6}$

   H  $\frac{1}{4}$

   J  $\frac{1}{2}$

Holt Mathematics

Name _____ Date _____ Class _____

## LESSON 5-10 Reading Strategies
### Compare and Contrast

When you compare two or more things, you look at how they are alike and how they are different. Equations with fractions follow the same rules as equations with whole numbers. Compare the whole number equation with the fraction equation.

|  | **Whole Number Equation** | **Fraction Equation** |  |
|---|---|---|---|
| **Step 1:** Divide both sides of the equation by the same number to get the variable by itself. | $3y = 36$ | $\frac{3}{4}y = 6$ |  |
| **Step 2:** Divide on both sides of the equation. | $\frac{3}{3}y = \frac{36}{3}$ | $\frac{4}{3} \cdot \frac{3}{4}y = 6 \cdot \frac{4}{3}$ <br> $y = 6 \cdot \frac{4}{3}$ <br> $y = \frac{24}{3}$ | ← Dividing by $\frac{3}{4}$ is the same as multiplying by $\frac{4}{3}$. <br> Multiply 6 by $\frac{4}{3}$. |
| **Step 3:** Simplify. | $y = 12$ | $y = 8$ |  |

**Compare the equations and answer each question.**

1. What is the same about solving equations with whole numbers and solving equations with fractions?

   _____

2. What is different about solving the two kinds of equations?

   _____

3. What is the first step to solve the problem $\frac{2}{5}w = 12$?

   _____

4. What is the second step?

   _____

5. The value of w is

   _____

Copyright © by Holt, Rinehart and Winston.
All rights reserved.

Holt Mathematics

Name _____ Date _____ Class _____

## Puzzles, Twisters & Teasers
**LESSON 5-10** *It's All Just Words*

Write these sentences as equations, solve the equations, and circle the answers. Put the letter of the correct answer on the spaces with the problem numbers. You will get 3 words that mean different things, but are all pronounced the same!

1. A number *n* is multiplied by one-third, giving a product of twelve.

   $n =$    4  **B**    36  **E**    15  **L**

2. Solve: A number *w* times one-third equals one-sixth.

   $w =$    $\frac{1}{3}$ **P**    2 **J**    $\frac{1}{2}$ **O**

3. Solve: The sum of *p* and three-sevenths gives one.

   $p =$    $\frac{4}{7}$ **U**    1 **D**    $\frac{1}{7}$ **A**

4. Solve: One-seventh times a number *t* equals one.

   $t =$    $\frac{1}{7}$ **G**    1 **N**    7 **F**

5. Subtracting three-fourths from a number *b* yields one and one-half.

   $b =$    $\frac{3}{4}$ **S**    $\frac{9}{4}$ **R**    $\frac{1}{2}$ **H**

This word is a number.    __ __ __ __
                          4  2  3  5

This word is a preposition.    __ __ __
                               4  2  5

This word is used in the game of golf.    __ __ __ __
                                          4  2  5  1

## Practice A
### 5-1 Least Common Multiple

**List the first five multiples.**

1. 2
   2, 4, 6, 8, 10
2. 6
   6, 12, 18, 24, 30
3. 12
   12, 24, 36, 48, 60
4. 3
   3, 6, 9, 12, 15
5. 7
   7, 14, 21, 28, 35
6. 10
   10, 20, 30, 40, 50

**Find the least common multiple (LCM).**

7. 2 and 3
   2: __
   3: __
   **6**
8. 2 and 8
   2: __
   8: __
   **8**
9. 2 and 4
   2: __
   4: __
   **4**
10. 2, 3, and 4
    2: __
    3: __
    4: __
    **12**
11. 3, 4, and 6
    3: __
    4: __
    6: __
    **12**
12. 3, 5, and 10
    3: __
    5: __
    10: __
    **30**
13. 2, 4, and 5
    **20**
14. 2, 4, and 6
    **12**
15. 2, 3, and 6
    **6**

16. Hot dogs come in packs of 8. Hot dog rolls come in packs of 12. What is the least number of packs of each Shawn should buy to have enough to serve 24 people and have none left over?

    **3 packs of hot dogs and 2 packs of rolls**

17. Debbie wants to invite 60 people to her party. Invitations come in packs of 12 and stamps come in sheets of 10. What is the least number of each she should buy to mail an invitation to each person and have no supplies left over?

    **5 packs of invitations and 6 sheets of stamps**

---

## Practice B
### 5-1 Least Common Multiple

**Find the least common multiple (LCM).**

1. 2 and 5 — **10**
2. 4 and 3 — **12**
3. 6 and 4 — **12**
4. 6 and 8 — **24**
5. 5 and 9 — **45**
6. 4 and 5 — **20**
7. 10 and 15 — **30**
8. 8 and 12 — **24**
9. 6 and 10 — **30**
10. 3, 6, and 9 — **18**
11. 2, 5, and 10 — **10**
12. 4, 7, and 14 — **28**
13. 3, 5, and 9 — **45**
14. 2, 5, and 8 — **40**
15. 3, 9, and 12 — **36**

16. Mr. Stevenson is ordering shirts and hats for his Boy Scout troop. There are 45 scouts in the troop. Hats come in packs of 3, and shirts come in packs of 5. What is the least number of packs of each he should order to so that each scout will have 1 hat and 1 shirt, and none will be left over?

    **15 packs of hats and 9 packs of shirts**

17. Tony wants to make 36 party bags. Glitter pens come in packs of 6. Stickers come in sheets of 4, and balls come in packs of 3. What is the least number of each package he should buy to have 1 of each item in every party bag, and no supplies left over?

    **6 packs of pens, 9 sheets of stickers, and 12 packs of balls**

18. Glenda is making 30 school supply baskets. Notepads come in packs of 5. Erasers come in packs of 15, and markers come in packs of 3. What is the least number of each package she should buy to have 1 of each item in every basket, and no supplies left over?

    **6 packs of notepads, 2 packs of erasers, and 10 packs of markers**

---

## Practice C
### 5-1 Least Common Multiple

**Find the least common multiple (LCM).**

1. 6 and 9 — **18**
2. 6 and 10 — **30**
3. 12 and 8 — **24**
4. 5 and 13 — **65**
5. 9 and 12 — **36**
6. 11 and 12 — **132**
7. 4, 7, and 14 — **28**
8. 5, 12, and 15 — **60**
9. 8, 14, and 16 — **112**
10. 6, 8, and 16 — **48**
11. 4, 8, and 64 — **64**
12. 6, 10, and 12 — **60**
13. 3, 6, 9, and 12 — **36**
14. 4, 6, 8, and 10 — **120**
15. 2, 6, 8, and 12 — **24**

16. Mr. Simon wants to make packages of art supplies for his students. Pads of paper come 4 to a box, pencils come 27 to a box, and erasers come 12 to a box. What is the least number of kits he can make if he wants each kit to be the same and he wants no supplies left over? How many boxes of paper must he buy? how many boxes of pencils? how many boxes of erasers?

    **108 kits; 27 boxes of paper; 4 boxes of pencils; 9 boxes of erasers**

17. Find the LCM and the GCF of 48 and 72. Now find the product of 48 and 72 and the product of the GCF and LCM. Describe the relationship between the two products. This relationship is true for all whole numbers. How could you use this relationship to solve problems?

    **The product of the two numbers is equal to the product of the LCM and the GCF. Possible answer: If you know the GCF of two numbers, you can divide the product of those two numbers by the GCF to find the LCM.**

    LCM = 144   GCF = 24

---

## Reteach
### 5-1 Least Common Multiple

The smallest number that is a multiple of two or more numbers is called the least common multiple (LCM).

To find the least common multiple of 3, 6, and 8, list the multiples for each number and put a circle around the LCM in the three lists.

Multiples of 3: 3, 6, 9, 12, 15, 18, 21, (24)
Multiples of 6: 6, 12, 18, (24), 30, 36, 42
Multiples of 8: 8, 16, (24), 32, 40, 48, 56
So 24 is the LCM of 3, 6, and 8.

**List the multiples of each number to help you find the least common multiple of each group.**

1. 3 and 4
   Multiples of 3: __
   Multiples of 4: __
   LCM: **12**
2. 5 and 7
   Multiples of 5: __
   Multiples of 7: __
   LCM: **35**
3. 8 and 12
   Multiples of 8: __
   Multiples of 12: __
   LCM: **24**
4. 2 and 9
   Multiples of 2: __
   Multiples of 9: __
   LCM: **18**
5. 4 and 6
   Multiples of 4: __
   Multiples of 6: __
   LCM: **12**
6. 4 and 10
   Multiples of 4: __
   Multiples of 10: __
   LCM: **20**
7. 2, 5, and 6
   Multiples of 2: __
   Multiples of 5: __
   Multiples of 6: __
   LCM: **30**
8. 3, 4, and 9
   Multiples of 3: __
   Multiples of 4: __
   Multiples of 9: __
   LCM: **36**
9. 8, 10, and 12
   Multiples of 8: __
   Multiples of 10: __
   Multiples of 12: __
   LCM: **120**

## Challenge
### 5-1 Moons Over Neptune

We measure one month by our moon's orbital period, or the time it takes the Moon to travel once around Earth, which is about 30 days. But what if you lived on Neptune? It has 8 moons! How could you pick just one moon to measure your months? One possible solution is to calculate one month based on when two of Neptune's moons are in conjunction at some arbitrary starting point in the sky, or appear to be in the same place in the sky. The diagram below shows some of the moons you could use to measure your months on Neptune.

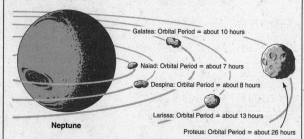

Galatea: Orbital Period = about 10 hours
Naiad: Orbital Period = about 7 hours
Despina: Orbital Period = about 8 hours
Larissa: Orbital Period = about 13 hours
Proteus: Orbital Period = about 26 hours

Use the diagram and least common multiples to complete the chart below. For each row, write how long your month on Neptune would be if you used those moons in conjunction as the length of one month.

| Neptune Moons to Use | Length of One Neptune Month |
|---|---|
| Naiad and Despina | about 56 hours |
| Larissa and Proteus | about 26 hours |
| Galatea and Despina | about 40 hours |
| Despina and Proteus | about 104 hours |

## Problem Solving
### 5-1 Least Common Multiple

Use the table to answer the questions.

| Party Supplies | |
|---|---|
| Item | Number per Pack |
| Invitations | 12 |
| Balloons | 30 |
| Paper plates | 10 |
| Paper napkins | 24 |
| Plastic cups | 15 |
| Noise makers | 5 |

1. You want to have an equal number of plastic cups and paper plates. What is the least number of packs of each you can buy?

   __3 packs of plates and__
   __2 packs of cups__

2. You want to invite 48 people to a party. What is the least number of packs of invitations and napkins you should buy to have one for each person and none left over?

   __4 packs of invitations and__
   __2 packs of napkins__

Circle the letter of the correct answer.

3. You want to have an equal number of noisemakers and balloons at your party. What is the least number of packs of each you can buy?
   A  1 pack of balloons and 1 pack of noise makers
   B  1 pack of balloons and 2 packs of noise makers
   **C** 1 pack of balloons and 6 packs of noise makers
   D  6 packs of balloons and 1 pack of noise makers

4. You bought an equal number of packs of plates and cups so that each of your 20 guests would have 3 cups and 2 plates. How many packs of each item did you buy?
   F  1 pack of cups and 1 pack of plates
   G  3 packs of cups and 4 packs of plates
   H  4 packs of cups and 3 packs of plates
   **J** 4 packs of cups and 4 packs of plates

5. The LCM for three items listed in the table is 60 packs. Which of the following are those three items?
   A  balloons, plates, noise makers
   **B** noise makers, invitations, balloons
   C  napkins, cups, plates
   D  balloons, napkins, plates

6. To have one of each item for 120 party guests, you buy 10 packs of one item and 24 packs of the other. What are those two items?
   F  plates and invitations
   G  balloons and cups
   H  napkins and plates
   **J** invitations and noise makers

## Reading Strategies
### 5-1 Understanding Vocabulary

**Least** means the smallest in size. The person with the least amount of homework has the smallest amount of work to do.

**Common** means shared. You may have classes in common with some of your friends.

A **multiple** is the answer to a multiplication problem.
The multiples of 5 are the answers to multiplying numbers by 5.
$1 \times 5 = 5$   $2 \times 5 = 10$   $3 \times 5 = 15$   $4 \times 5 = 20$

The **least common multiple** is the smallest multiple two numbers have in common.

Follow the steps for finding the least common multiple of 5 and 10.

1. List the first 10 multiples of 5.
   __5, 10, 15, 20, 25, 30, 35, 40, 45, 50__

2. List the first 5 multiples of 10.
   __10, 20, 30, 40, 50__

3. What multiples do 5 and 10 have in common?
   __10, 20, 30, 40, 50__

4. Write the smallest multiple that 5 and 10 have in common. __10__

5. What is the least common multiple of 5 and 10? __10__

6. To find the least common multiple of two numbers, what is the first thing you should do?
   __List the multiples of both numbers.__

7. What should you do next?
   __Compare the multiples they have in common.__

8. How do you know which of the common multiples is the least common multiple?
   __It is the smallest multiple.__

## Puzzles, Twisters & Teasers
### 5-1 Math Abbreviation

Draw a line from each pair of numbers to common multiples for the numbers. Sometimes you will need to draw two lines from the same pair of numbers.

When you have finished, you will see a famous math abbreviation.

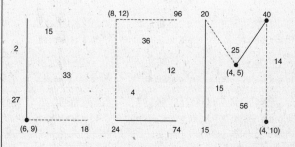

## Practice A
### 5-2 Adding and Subtracting with Unlike Denominators

Write the least common denominator for each pair of fractions.

1. $\frac{1}{2}, \frac{2}{4}$     2. $\frac{1}{8}, \frac{2}{3}$     3. $\frac{1}{6}, \frac{1}{4}$
   4                              24                                12

4. $\frac{1}{3}, \frac{1}{5}$     5. $\frac{1}{5}, \frac{3}{4}$     6. $\frac{1}{5}, \frac{7}{10}$
   15                             20                                10

Add or subtract. Write each answer in simplest form.

7. $\frac{1}{2} + \frac{2}{3}$     8. $\frac{1}{2} - \frac{1}{4}$     9. $\frac{3}{4} - \frac{2}{3}$
   $1\frac{1}{6}$                  $\frac{1}{4}$                      $\frac{1}{12}$

10. $\frac{2}{5} - \frac{1}{10}$   11. $\frac{1}{6} + \frac{1}{3}$    12. $\frac{1}{5} + \frac{7}{10}$
    $\frac{3}{10}$                  $\frac{1}{2}$                       $\frac{9}{10}$

13. $\frac{5}{8} - \frac{1}{4}$    14. $\frac{1}{5} + \frac{1}{4}$    15. $\frac{1}{2} - \frac{3}{8}$
    $\frac{3}{8}$                   $\frac{9}{20}$                      $\frac{1}{8}$

16. $\frac{2}{7} - \frac{1}{14}$   17. $\frac{3}{5} + \frac{1}{15}$   18. $\frac{5}{6} + \frac{1}{2}$
    $\frac{3}{14}$                  $\frac{2}{3}$                       $1\frac{1}{3}$

19. Alice practices the piano $\frac{3}{4}$ hour every day. Today, however, she practiced for $\frac{1}{2}$ hour longer than usual. How long did Alice practice the piano today?

    $1\frac{1}{4}$ hours

20. One lap around the school's track is $\frac{1}{4}$ mile. Tyler ran two times around the track. Then he ran $\frac{5}{6}$ mile home. How far did Tyler run in all?

    $1\frac{1}{3}$ miles

/20

## Practice B
### 5-2 Adding and Subtracting with Unlike Denominators

Add or subtract. Write each answer in simplest form.

1. $\frac{6}{7} + \frac{1}{3}$     2. $\frac{3}{7} - \frac{2}{5}$     3. $\frac{1}{4} + \frac{3}{8}$
   $1\frac{4}{21}$                 $\frac{1}{35}$                     $\frac{5}{8}$

4. $\frac{7}{8} - \frac{2}{3}$     5. $\frac{1}{6} + \frac{3}{5}$     6. $\frac{5}{6} - \frac{2}{3}$
   $\frac{5}{24}$                  $\frac{23}{30}$                    $\frac{1}{6}$

7. $\frac{5}{9} - \frac{1}{3}$     8. $\frac{7}{8} + \frac{3}{4}$     9. $\frac{5}{12} - \frac{1}{6}$
   $\frac{2}{9}$                   $1\frac{5}{8}$                     $\frac{1}{4}$

10. $\frac{4}{5} - \frac{7}{11}$   11. $\frac{4}{9} + \frac{5}{6}$    12. $\frac{5}{8} + \frac{2}{3}$
    $\frac{9}{55}$                  $1\frac{5}{18}$                    $1\frac{7}{24}$

Evaluate each expression for $b = \frac{1}{3}$. Write your answer in simplest form.

13. $b + \frac{5}{8}$              14. $\frac{7}{9} - b$              15. $\frac{2}{7} + b$
    $\frac{23}{24}$                 $\frac{4}{9}$                      $\frac{13}{21}$

16. $b + b$                        17. $\frac{11}{12} - b$            18. $\frac{3}{4} - b$
    $\frac{2}{3}$                   $\frac{7}{12}$                     $\frac{5}{12}$

19. There are three grades in Kyle's middle school—sixth, seventh, and eighth. One-third of the students are in sixth grade and $\frac{1}{4}$ are in seventh grade. What fraction of the schools' students are in eighth grade?

    $\frac{5}{12}$ of the students

20. Sarah is making a dessert that calls for $\frac{4}{5}$ cup of crushed cookies. If she has already crushed $\frac{7}{10}$ cup, how much more does she need?

    $\frac{1}{10}$ cup

/20

## Practice C
### 5-2 Adding and Subtracting with Unlike Denominators

Evaluate. Write each answer in simplest form.

1. $\frac{11}{12} + \frac{3}{5}$   2. $\frac{7}{12} - \frac{5}{16}$   3. $\frac{5}{6} + \frac{3}{10}$
   $1\frac{31}{60}$                $\frac{13}{48}$                    $1\frac{2}{15}$

4. $\frac{3}{4} - \frac{3}{14}$    5. $\frac{1}{2} + \frac{5}{17}$    6. $\frac{4}{5} - \frac{2}{9}$
   $\frac{15}{28}$                 $\frac{27}{34}$                    $\frac{26}{45}$

7. $\frac{7}{8} - \frac{5}{12}$    8. $\frac{3}{16} + \frac{5}{6}$    9. $\frac{3}{16} + \frac{5}{32}$
   $\frac{11}{24}$                 $1\frac{1}{48}$                    $\frac{11}{32}$

10. $\frac{11}{12} - \frac{4}{9} + \frac{1}{2}$   11. $\frac{2}{15} + \frac{7}{25} - \frac{2}{5}$   12. $\frac{3}{14} - \frac{1}{8} + \frac{4}{7}$
    $\frac{35}{36}$                                $\frac{1}{75}$                                    $\frac{37}{56}$

Evaluate each expression for $b = \frac{2}{5}$. Write your answer in simplest form.

13. $b + \frac{9}{14}$             14. $\frac{7}{12} - b$             15. $\frac{11}{16} - b$
    $1\frac{3}{70}$                 $\frac{11}{60}$                    $\frac{23}{80}$

16. $b + \frac{8}{11}$             17. $\frac{4}{7} - b$              18. $\frac{14}{15} + b$
    $1\frac{7}{55}$                 $\frac{6}{35}$                     $1\frac{1}{3}$

19. Ben, Shaneeka, and Phil live on the same street. Ben lives $\frac{6}{11}$ mile north of Phil, and $\frac{1}{3}$ mile north of Shaneeka. How far does Shaneeka live from Phil?

    $\frac{7}{33}$ of a mile

20. At the frog-jumping contest, Trevor's frog jumped $\frac{5}{6}$ foot and then $\frac{5}{9}$ foot. Mei's frog jumped $\frac{4}{5}$ foot and then $\frac{1}{2}$ foot. Whose frog jumped the farthest in all? How much farther?

    Trevor's frog; $\frac{4}{45}$ foot farther

## Reteach
### 5-2 Adding and Subtracting with Unlike Denominators

Unlike fractions have different denominators. To add and subtract fractions, you must have a common denominator. The least common denominator (LCD) is the least common multiple of the denominators.

To add or subtract unlike fractions, first find the LCD of the fractions.

$\frac{2}{3} + \frac{1}{4}$

Multiples of 4: 4, 8, **12**,…
Multiples of 3: 3, 6, 9, **12**,…
The LCD is 12.

Next, use fraction strips to find equivalent fractions.

Then use fraction strips to find the sum or difference.

$\frac{8}{12} + \frac{3}{12} = \frac{11}{12}$
So, $\frac{2}{3} + \frac{1}{4} = \frac{11}{12}$

Use fraction strips to find each sum or difference. Write your answer in simplest form.

1. $\frac{1}{4} + \frac{1}{8}$   2. $\frac{5}{6} - \frac{2}{3}$   3. $\frac{3}{4} - \frac{1}{3}$   4. $\frac{3}{5} + \frac{3}{10}$
   $\frac{3}{8}$                 $\frac{1}{6}$                    $\frac{5}{12}$                   $\frac{9}{10}$

5. $\frac{3}{4} + \frac{1}{6}$   6. $\frac{1}{2} + \frac{3}{8}$   7. $\frac{2}{3} - \frac{1}{6}$   8. $\frac{1}{3} - \frac{1}{4}$
   $\frac{11}{12}$               $\frac{7}{8}$                    $\frac{1}{2}$                    $\frac{1}{12}$

/18

85        Holt Mathematics

## LESSON 5-2 Challenge
### Egyptian Fractions

Did you know that ancient Egyptians used fractions 5,000 years ago? Some of their fractions were like the ones we use today. However, the Egyptians only used **unit fractions**, or fractions with a numerator of 1. All other fractions had to be written as a sum of unit fractions. And no sum could repeat the same unit fraction! For example, the Egyptians would write $\frac{3}{4}$ as $\frac{1}{2} + \frac{1}{4}$. They would not write $\frac{1}{4} + \frac{1}{4} + \frac{1}{4}$.

Ancient Egyptians did not have paper. They recorded their math work on papyrus, or thin strips of dried plants. Study the Egyptian fractions recorded on the papyrus scrolls below. Then write each fraction the way we do today.

1. 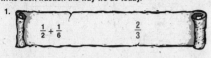   $\frac{2}{3}$

2.    $\frac{4}{5}$

3.    $\frac{5}{6}$

4.    $\frac{3}{8}$

5.    $\frac{9}{10}$

---

## LESSON 5-2 Problem Solving
### Adding and Subtracting with Unlike Denominators

Use the circle graph to answer the questions. Write each answer in simplest form.

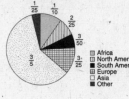

World Population, 2001
- Africa
- North America
- South America
- Europe
- Asia
- Other

1. On which two continents do most people live? How much of the total population do they make up together?

   Asia and Europe; $\frac{18}{25}$ of the population

2. How much of the world's population live in either North America or South America?

   $\frac{7}{50}$ of the population

3. How much more of the world's total population lives in Asia than in Africa?

   $\frac{1}{2}$ of the population

Circle the letter of the correct answer.

4. How much of Earth's total population do people in Asia and Africa make up all together?
   - A $\frac{3}{10}$ of the population
   - B $\frac{2}{5}$ of the population
   - **C** $\frac{7}{10}$ of the population
   - D $\frac{7}{5}$ of the population

5. What is the difference between North America's part of the total population and Africa's part?
   - **F** Africa has $\frac{1}{50}$ more.
   - G Africa has $\frac{1}{50}$ less.
   - H Africa has $\frac{9}{50}$ more.
   - J Africa has $\frac{9}{50}$ less.

6. How much more of the population lives in Europe than in North America?
   - **A** $\frac{1}{25}$ of the population
   - B $\frac{1}{5}$ of the population
   - C $\frac{1}{15}$ of the population
   - D $\frac{1}{10}$ of the population

7. How much of the world's population lives in North America and Europe?
   - F $\frac{1}{25}$ of the population
   - G $\frac{1}{15}$ of the population
   - **H** $\frac{1}{5}$ of the population
   - J $\frac{1}{20}$ of the population

---

## LESSON 5-2 Reading Strategies
### Use Fraction Bars

You can use fraction bars to show $\frac{1}{2}$ and $\frac{1}{3}$.

| $\frac{1}{2}$ | $\frac{1}{3}$ |

These fractions have denominators that are different. They are called **unlike fractions**.

To add or subtract unlike fractions, the denominators must be the same. They must have a **common denominator**.

| $\frac{1}{6}$ $\frac{1}{6}$ $\frac{1}{6}$ | $\frac{1}{6}$ $\frac{1}{6}$ |

The common denominator for $\frac{1}{2}$ and $\frac{1}{3}$ is 6.

To get a common denominator for two fractions, multiply the denominators.

halves • thirds = sixths, or 2 • 3 = 6

1. What are unlike fractions?

   <u>Fractions that have different denominators.</u>

2. If you want to add or subtract unlike fractions, what do you need to do?

   <u>Find a common denominator.</u>

3. How do you get a common denominator for $\frac{1}{2}$ and $\frac{1}{3}$?

   <u>Multiply the denominators.</u>

4. How many sixths are in one-half?

   <u>three</u>

5. How many sixths are in one-third?

   <u>two</u>

6. What is the sum of one-half and one-third?

7. What is the difference between one-half and one-third? $\frac{1}{6}$

---

## LESSON 5-2 Puzzles, Twisters & Teasers
### The Truth of the Matter

Decide whether each statement is true or false. Circle your answer.

Use the letters of your true answers and rearrange them to answer the question.

1. one-third plus one-seventh is greater than one-half

   M true            **E** false
   $\frac{1}{3} + \frac{1}{7} = \frac{7}{21} + \frac{3}{21} = \frac{10}{21}$

2. one-fourth plus one-ninth is greater than one-third

   **C** true            P false
   $\frac{1}{4} + \frac{1}{9} = \frac{9}{36} + \frac{4}{36} = \frac{13}{36}$

3. four-fifths minus one-third is less than one-half

   **D** true            N false
   $\frac{4}{5} - \frac{1}{3} = \frac{12}{15} - \frac{5}{15} = \frac{7}{15}$

4. three-fourths minus three tenths equals one-half

   A true            **S** false
   $\frac{3}{4} - \frac{3}{10} = \frac{15}{20} - \frac{6}{20} = \frac{9}{20}$

5. seven-twelfths plus three-eighths is greater than one

   J true            **T** false
   $\frac{7}{12} + \frac{3}{8} = \frac{14}{24} + \frac{9}{24} = \frac{23}{24}$

6. three-fourths minus three-twelfths equals one-half

   **L** true            B false
   $\frac{3}{4} - \frac{3}{12} = \frac{9}{12} - \frac{3}{12} = \frac{6}{12} = \frac{1}{2}$

What must you find in order to add or subtract unlike fractions?

The  L  C  D

## Practice A
### 5-3 Adding and Subtracting Mixed Numbers

Estimate each sum or difference to the nearest whole number.

1. $2\frac{1}{5} + 1\frac{1}{4}$    2. $3\frac{1}{6} + 1\frac{4}{5}$    3. $4\frac{1}{2} - 2\frac{1}{8}$

    3           5           2

4. $1\frac{1}{2} + 2\frac{3}{4}$    5. $2\frac{2}{3} - 1\frac{5}{6}$    6. $1\frac{1}{7} - 1\frac{1}{8}$

   4 or 5       1           0

Find each sum or difference. Write the answer in simplest form.

7. $1\frac{1}{2} + 3\frac{1}{4}$    8. $10\frac{3}{5} - 8\frac{1}{10}$    9. $3\frac{1}{4} + 2\frac{5}{6}$

   $4\frac{3}{4}$       $2\frac{1}{2}$       $6\frac{1}{12}$

10. $3\frac{2}{6} - 1\frac{1}{3}$    11. $10\frac{2}{3} - 9\frac{1}{4}$    12. $4\frac{2}{15} + 1\frac{1}{5}$

   2        $1\frac{5}{12}$       $5\frac{1}{3}$

13. $8\frac{1}{2} + 2\frac{1}{3}$    14. $12\frac{1}{2} - 10\frac{1}{8}$    15. $7\frac{7}{8} + 1\frac{1}{6}$

   $10\frac{5}{6}$      $2\frac{3}{8}$       $8\frac{7}{24}$

16. $2\frac{7}{12} + 1\frac{1}{8}$    17. $4\frac{1}{6} - 1\frac{1}{9}$    18. $3\frac{1}{7} + 3\frac{1}{3}$

   $3\frac{17}{24}$      $3\frac{1}{18}$      $6\frac{10}{21}$

19. $1\frac{2}{3} - 1\frac{1}{2}$    20. $5\frac{2}{5} + 1\frac{1}{2}$    21. $2\frac{2}{3} + 2\frac{1}{5}$

   $\frac{1}{6}$       $6\frac{9}{10}$      $4\frac{8}{15}$

22. Jack babysat for $4\frac{1}{4}$ hours on Friday night. He babysat for $3\frac{2}{3}$ hours on Saturday night. How many hours did he babysit in all?

   $7\frac{11}{12}$ hours

23. Bonita planted an oak tree and an elm tree in her backyard. Three years later, the oak tree was $5\frac{5}{12}$ feet tall, and the elm tree was $7\frac{1}{2}$ feet tall. How much taller was the elm tree?

   $2\frac{1}{3}$ feet

/23

---

## Practice B
### 5-3 Adding and Subtracting Mixed Numbers

Find each sum or difference. Write the answer in simplest form.

1. $4\frac{3}{8} + 5\frac{1}{4}$    2. $11\frac{2}{5} - 8\frac{1}{3}$    3. $7\frac{1}{3} + 3\frac{2}{9}$

   $9\frac{5}{8}$       $3\frac{1}{15}$      $10\frac{5}{9}$

4. $22\frac{5}{6} - 17\frac{1}{4}$    5. $32\frac{4}{7} - 14\frac{1}{3}$    6. $12\frac{1}{4} + 5\frac{1}{12}$

   $5\frac{7}{12}$     $18\frac{5}{21}$     $17\frac{1}{3}$

7. $29\frac{1}{3} - 14\frac{1}{6}$    8. $5\frac{3}{4} - 1\frac{7}{11}$    9. $21\frac{1}{6} + 1\frac{3}{8}$

   $15\frac{1}{6}$     $4\frac{5}{44}$     $22\frac{13}{24}$

10. $15\frac{7}{12} - 14\frac{3}{8}$    11. $5\frac{6}{15} + 4\frac{3}{10}$    12. $25\frac{1}{7} + 25\frac{2}{5}$

   $1\frac{5}{24}$     $9\frac{7}{10}$     $50\frac{19}{35}$

13. $3\frac{2}{5} + 1\frac{1}{3}$    14. $1\frac{2}{5} - 1\frac{2}{3}$    15. $3\frac{3}{5} - 2\frac{1}{2}$

   $4\frac{11}{15}$     $\frac{1}{5}$     $1\frac{1}{10}$

16. $6\frac{3}{4} - 3\frac{3}{10}$    17. $4\frac{4}{5} + 2\frac{1}{10}$    18. $32\frac{1}{2} + 5\frac{1}{3}$

   $3\frac{9}{20}$     $6\frac{9}{10}$     $37\frac{5}{6}$

19. Donald is making a party mix. He bought $2\frac{1}{4}$ pounds of pecans and $3\frac{1}{5}$ pounds of walnuts. How many pounds of nuts did Donald buy in all?

   $5\frac{9}{20}$ pounds

20. Mrs. Watson's cookie recipe calls for $3\frac{4}{7}$ cups of sugar. Mr. Clark's cookie recipe calls for $4\frac{2}{3}$ cups of sugar. How much more sugar does Mr. Clark's recipe use?

   $1\frac{2}{21}$ cups more

21. Tasha's cat weighs $15\frac{5}{12}$ lb. Naomi's cat weighs $11\frac{1}{3}$ lb. Can they bring both of their cats to the vet in a carrier that can hold up to 27 pounds? Explain.

Yes; because the cats' combined weight is $26\frac{3}{4}$ pounds, which is less the 27 pounds

---

## Practice C
### 5-3 Adding and Subtracting Mixed Numbers

Add or subtract. Write each answer in simplest form.

1. $8\frac{1}{10} + 2\frac{2}{25}$    2. $3\frac{1}{2} - 1\frac{10}{17}$    3. $2\frac{1}{6} - 1\frac{1}{16}$

   $10\frac{9}{50}$     $1\frac{31}{34}$     $1\frac{5}{48}$

4. $6\frac{5}{6} + 14\frac{1}{9}$    5. $11\frac{7}{8} - 5\frac{3}{12}$    6. $4\frac{2}{13} + 3\frac{1}{2}$

   $20\frac{17}{18}$     $6\frac{5}{8}$     $7\frac{17}{26}$

7. $2\frac{1}{12} - 1\frac{1}{14}$    8. $5\frac{7}{16} - 4\frac{2}{5}$    9. $1\frac{1}{8} + 1\frac{1}{9}$

   $1\frac{1}{84}$     $1\frac{3}{80}$     $2\frac{17}{72}$

Evaluate. Write each answer as a fraction in simplest form.

10. $17\frac{2}{7} + 1.6$    11. $8\frac{1}{5} + 1.5$    12. $19\frac{9}{25} - 6.3$

   $18\frac{31}{35}$     $9\frac{7}{10}$     $13\frac{3}{50}$

13. $23\frac{9}{10} - 18.7$    14. $11.42 + \frac{1}{25}$    15. $12\frac{17}{20} - 4.05$

   $5\frac{1}{5}$     $11\frac{23}{50}$     $8\frac{4}{5}$

16. Mattie's camping gear and food weighed $24\frac{7}{15}$ pounds at the beginning of the weekend. His food weighed $10\frac{2}{9}$ pounds. After Mattie ate all of his food, how much did his gear weigh?

   $14\frac{11}{45}$ pounds

17. Cindy and Tara are saving money to buy a present for their teacher that costs $37.50. So far, Cindy has saved $15\frac{1}{4}$ dollars, and Tara has saved $11\frac{4}{5}$ dollars. How much more do they need to buy the present?

   $10.45 more

18. Terry rode her bike for $15\frac{1}{6}$ miles last week. This week she rode her bike for $21\frac{1}{2}$ miles. How many miles did Terry ride her bike during these two weeks? How many more miles did she ride her bike during the second week?

   $36\frac{2}{3}$ miles; $6\frac{1}{3}$ miles

/19

---

## Reteach
### 5-3 Adding and Subtracting Mixed Numbers

You can use what you know about improper fractions to add and subtract mixed numbers.

To find the sum or difference of mixed numbers, first write the mixed numbers as improper fractions.

A. $3\frac{1}{4} + 2\frac{1}{3}$      B. $4\frac{1}{2} - 2\frac{2}{3}$

$= \frac{13}{4} + \frac{7}{3}$      $= \frac{9}{2} - \frac{8}{3}$

Next, find equivalent fractions with a least common denominator.

$\frac{13}{4} + \frac{7}{3}$      $\frac{9}{2} - \frac{8}{3}$

$= \frac{39}{12} + \frac{28}{12}$      $= \frac{27}{6} - \frac{16}{6}$

Then add or subtract the like fractions.

$\frac{39}{12} + \frac{28}{12}$      $\frac{27}{6} - \frac{16}{6}$

$= \frac{67}{12}$      $= \frac{11}{6}$

Write the answer as a mixed number in simplest form.

$\frac{67}{12}$      $\frac{11}{6}$

$= 5\frac{7}{12}$      $= 1\frac{5}{6}$

So, $3\frac{1}{4} + 2\frac{1}{3} = 5\frac{7}{12}$.      So, $4\frac{1}{2} - 2\frac{2}{3} = 1\frac{5}{6}$.

Find each sum or difference. Write your answer in simplest form.

1. $1\frac{1}{4} + 1\frac{1}{2}$    2. $3\frac{1}{6} + 1\frac{2}{3}$    3. $2\frac{1}{8} + 4\frac{1}{2}$    4. $4\frac{1}{3} + 1\frac{1}{2}$

$= \frac{5}{4} + \frac{3}{2}$    $= \frac{19}{6} + \frac{5}{3}$    $= \frac{17}{8} + \frac{9}{2}$    $= \frac{13}{3} + \frac{3}{2}$

$= \frac{5}{4} + \frac{6}{4}$    $= \frac{19}{6} + \frac{10}{6}$    $= \frac{17}{8} + \frac{36}{8}$    $= \frac{26}{6} + \frac{9}{6}$

   $2\frac{3}{4}$       $4\frac{5}{6}$       $6\frac{5}{8}$       $5\frac{5}{6}$

5. $2\frac{3}{5} + 1\frac{1}{10}$    6. $3\frac{1}{6} + 1\frac{1}{12}$    7. $2\frac{5}{8} - 1\frac{1}{4}$    8. $5\frac{2}{3} - 2\frac{1}{4}$

   $3\frac{7}{10}$       $4\frac{1}{4}$       $1\frac{3}{8}$       $3\frac{5}{12}$

/24

87    **Holt Mathematics**

## LESSON 5-3 Challenge
### Maximum Snakes

The bar graph below shows the maximum lengths for the longest snakes in the world. Use the graph to find how much each of the snakes in the City Zoo is below its maximum length.

**World's Longest Snakes**

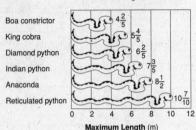

- Boa constrictor: $4\frac{2}{5}$
- King cobra: $5\frac{4}{5}$
- Diamond python: $6\frac{5}{6}$
- Indian python: $7\frac{3}{5}$
- Anaconda: $8\frac{1}{2}$
- Reticulated python: $10\frac{7}{10}$

Maximum Length (m)

**Snakes in the City Zoo**

| Snake | Length (in meters) | Difference from Maximum Length |
|---|---|---|
| Kevin (king cobra) | $3\frac{1}{2}$ | $2\frac{3}{10}$ meters |
| Annie (anaconda) | $5\frac{1}{3}$ | $3\frac{1}{6}$ meters |
| Bob (boa constrictor) | $3\frac{1}{4}$ | $1\frac{3}{20}$ meters |
| Ivy (Indian python) | $4\frac{2}{7}$ | $3\frac{11}{35}$ meters |
| Reggie (reticulated python) | $8\frac{3}{5}$ | $2\frac{1}{10}$ meters |
| Diana (diamond python) | $4\frac{3}{8}$ | $2\frac{1}{40}$ meters |

---

## LESSON 5-3 Problem Solving
### Adding and Subtracting Mixed Numbers

Write the correct answer in simplest form.

1. Of the planets in our solar system, Jupiter and Neptune have the greatest surface gravity. Jupiter's gravitational pull is $2\frac{16}{25}$ stronger than Earth's, and Neptune's is $1\frac{1}{5}$ stronger. What is the difference between Jupiter's and Neptune's surface gravity levels?

   **Jupiter's is $1\frac{11}{25}$ higher.**

2. Escape velocity is the speed a rocket must attain to overcome a planet's gravitational pull. Earth's escape velocity is $6\frac{9}{10}$ miles per second! The Moon's escape velocity is $5\frac{2}{5}$ miles per second slower. How fast does a rocket have to launch to escape the moon's gravity?

   **$1\frac{1}{2}$ miles per second**

3. The two longest total solar eclipses occurred in 1991 and 1992. The first one lasted $6\frac{5}{6}$ minutes. The eclipse of 1992 lasted $5\frac{1}{3}$ minutes. How much longer was 1991's eclipse?

   **$1\frac{1}{2}$ minutes**

4. The two largest meteorites found in the U.S. landed in Canyon Diablo, Arizona, and Willamette, Oregon. The Arizona meteorite weighs $33\frac{1}{10}$ tons! Oregon's weighs $16\frac{1}{2}$ tons. How much do the two meteorites weigh in all?

   **$49\frac{3}{5}$ tons**

Circle the letter of the correct answer.

5. Not including the Sun, Proxima Centauri is the closest star to Earth. It is $4\frac{11}{6}$ light years away! The next closest star is Alpha Centauri. It is $\frac{13}{100}$ light years farther than Proxima. How far is Alpha Centauri from Earth?

   **A** $4\frac{7}{20}$ light years
   B $4\frac{13}{100}$ light years
   C $4\frac{6}{25}$ light years
   D $4\frac{1}{50}$ light years

6. It takes about $5\frac{1}{3}$ minutes for light from the Sun to reach Earth. The Moon is closer to Earth, so its light reaches Earth faster—about $5\frac{19}{60}$ minutes faster than from the Sun. How long does light from the Moon take to reach Earth?

   F $\frac{3}{10}$ of a minute
   **G** $\frac{1}{60}$ of a minute
   H $\frac{1}{3}$ of a minute
   J $\frac{4}{15}$ of a minute

---

## LESSON 5-3 Reading Strategies
### Summarize

$\frac{2}{3} + \frac{1}{4}$

To add $\frac{2}{3} + \frac{1}{4}$, the denominators must be the same.

**First,** multiply the denominators to get a common denominator.

$3 \times 4 = 12$

1. How do you get a common denominator for two fractions?

   **Multiply the denominators of the fractions.**

2. What is the common denominator of $\frac{2}{3}$ and $\frac{1}{4}$? **12**

**Next,** change both fractions into equivalent fractions with 12 as the denominator.

$\frac{2}{3} = \frac{8}{12}$ (×4)     $\frac{1}{4} = \frac{3}{12}$ (×3)

**Finally,** add the fractions together.

Look at the drawing above to answer each question.

3. What equivalent fraction is formed when $\frac{2}{3}$ is changed into twelfths?

   $\frac{8}{12}$

4. What equivalent fraction is formed when $\frac{1}{4}$ is changed into twelfths?

   $\frac{3}{12}$

5. What is $\frac{8}{12} + \frac{3}{12}$? $\frac{11}{12}$

6. Write how you would add two fractions that have different denominators.

   **Possible answer: Find a common denominator, find equivalent fractions for each, and add.**

---

## LESSON 5-3 Puzzles, Twisters & Teasers
### Work Before Play

**Down**

1. The whole number part of the answer: $7\frac{5}{6} - 2\frac{1}{2}$.
2. When you have ___ denominators,
4. you must find the ___.
6. The common denominator of $2\frac{4}{5} + 3\frac{1}{2}$.

**Across**

1. Solve for $n$.  $\frac{n}{3} + 2\frac{1}{3} = 3\frac{2}{3}$
3. $\frac{7}{3}$ is the ___ of $\frac{3}{7}$.
5. The fractional part of the answer: $7\frac{5}{6} - 2\frac{1}{2}$.
7. The numerator of $\frac{7}{8}$.

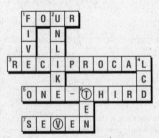

```
¹F O ²U R
    I   N
    V   L
³R E C I P R O C A ⁴L
    K           C
⁵O N E - ⁶T H I R D
            E
    ⁵S E ⁶V E N
```

The circled letters in the crossword will give you the answer.

What two letters can keep you from doing your homework? **T V**

## Practice A
### 5-4 Regrouping to Subtract Mixed Numbers

Regroup each mixed number by regrouping a 1 from the whole number.

1. $1\frac{1}{4}$    $\frac{5}{4}$
2. $8\frac{5}{12}$    $7\frac{17}{12}$
3. $4\frac{5}{9}$    $3\frac{14}{9}$
4. $2\frac{1}{3}$    $1\frac{4}{3}$
5. $7\frac{1}{9}$    $6\frac{10}{9}$
6. $10\frac{3}{7}$    $9\frac{10}{7}$

Subtract. Write each answer in simplest form.

7. $2 - \frac{2}{3}$    $1\frac{1}{3}$
8. $1 - \frac{1}{4}$    $\frac{3}{4}$
9. $5\frac{1}{4} - 3\frac{1}{2}$    $1\frac{3}{4}$
10. $2\frac{1}{3} - 1\frac{5}{6}$    $\frac{1}{2}$
11. $1\frac{4}{9} - \frac{2}{3}$    $\frac{7}{9}$
12. $2\frac{1}{4} - 1\frac{7}{8}$    $\frac{3}{8}$
13. $5\frac{3}{10} - 1\frac{4}{5}$    $3\frac{1}{2}$
14. $2\frac{1}{4} - \frac{11}{16}$    $1\frac{9}{16}$
15. $3\frac{1}{3} - 2\frac{4}{5}$    $\frac{8}{15}$

16. At the pie-eating contest, Dina ate $3\frac{1}{3}$ pies. Mason ate $2\frac{5}{6}$ pies. How much more pie did Dina eat than Mason?    $\frac{1}{2}$ of a pie

17. When Latoya bought her angel fish, it was $1\frac{1}{2}$ inches long. Now it is $2\frac{1}{3}$ inches long. How much did her angel fish grow?    $\frac{5}{6}$ of an inch

## Practice B
### 5-4 Regrouping to Subtract Mixed Numbers

Subtract. Write each answer in simplest form.

1. $4 - 2\frac{3}{8}$    $1\frac{5}{8}$
2. $5\frac{1}{6} - 2\frac{2}{3}$    $2\frac{1}{2}$
3. $14 - 8\frac{2}{9}$    $5\frac{7}{9}$
4. $19\frac{1}{7} - 5\frac{1}{3}$    $13\frac{17}{21}$
5. $7\frac{1}{4} - 3\frac{5}{8}$    $3\frac{5}{8}$
6. $10\frac{1}{5} - 5\frac{7}{10}$    $4\frac{1}{2}$
7. $1\frac{1}{6} - \frac{7}{9}$    $\frac{7}{18}$
8. $9\frac{1}{4} - 1\frac{7}{16}$    $7\frac{13}{16}$
9. $6\frac{1}{5} - 3\frac{1}{4}$    $2\frac{19}{20}$

Evaluate each expression for $a = 1\frac{1}{2}$, $b = 2\frac{1}{3}$, $c = \frac{1}{4}$, and $d = 3$. Write the answer in simplest form.

10. $b - a$    $\frac{5}{6}$
11. $a - c$    $1\frac{1}{4}$
12. $b - c$    $2\frac{1}{12}$
13. $d - a$    $1\frac{1}{2}$
14. $d - b$    $\frac{2}{3}$
15. $d - c$    $2\frac{3}{4}$

16. Tim had 6 feet of wrapping paper for Kylie's birthday present. He used $3\frac{3}{8}$ feet of the paper to wrap her gift. How much paper did Tim have left?    $2\frac{5}{8}$ feet of paper

17. At his last doctor's visit, Pablo was $60\frac{1}{2}$ inches tall. At today's visit, he measured $61\frac{1}{6}$ inches. How much did Pablo grow between visits?    $\frac{2}{3}$ inch

18. Yesterday, Danielle rode her bike for $5\frac{1}{2}$ miles. Today, she rode her bike for $6\frac{1}{4}$ miles. How much farther did Danielle ride her bike today?    $\frac{3}{4}$ mile

## Practice C
### 5-4 Regrouping to Subtract Mixed Numbers

Subtract. Write each answer in simplest form.

1. $7 - 3\frac{11}{12}$    $3\frac{1}{12}$
2. $8\frac{4}{13} - 1\frac{19}{26}$    $6\frac{15}{26}$
3. $14\frac{5}{12} - 3\frac{7}{8}$    $10\frac{13}{24}$
4. $5\frac{1}{7} - 2\frac{2}{3}$    $2\frac{10}{21}$
5. $19\frac{1}{12} - 4\frac{4}{9}$    $14\frac{23}{36}$
6. $19\frac{3}{5} - 6\frac{5}{7}$    $12\frac{31}{35}$
7. $17\frac{1}{14} - 8\frac{7}{8}$    $8\frac{11}{56}$
8. $14\frac{2}{7} - 11\frac{3}{4}$    $2\frac{15}{28}$
9. $22\frac{3}{11} - 2\frac{9}{10}$    $19\frac{41}{110}$

Evaluate each expression for $a = 6\frac{3}{8}$, $b = 5\frac{1}{6}$, $c = 7\frac{1}{2}$, and $d = 10$. Write the answer in simplest form.

10. $a - b$    $1\frac{5}{24}$
11. $c - a$    $\frac{17}{24}$
12. $c - b$    $1\frac{11}{12}$
13. $d - a$    $3\frac{5}{8}$
14. $d - b$    $4\frac{5}{6}$
15. $d - c$    $2\frac{11}{12}$

16. Annie bought $21\frac{2}{5}$ pounds of clay. She used $15\frac{5}{6}$ pounds of the clay to make a vase, and $1\frac{4}{5}$ pounds to make a coaster. How much clay does she have left?    $3\frac{23}{30}$ pounds

17. In January, a chef bought $15\frac{1}{8}$ pounds of ground beef. In February, he bought $3\frac{4}{5}$ pounds less. Then in March he bought $1\frac{19}{20}$ pounds less than in February. How many pounds of ground beef did the chef buy in March?    $9\frac{3}{8}$ pounds

18. John is training for the triathlon. He wants to cover a distance of $16\frac{1}{4}$ miles today. If he runs for $6\frac{1}{2}$ miles and rides his bike for $7\frac{4}{5}$ miles, how far does he have to swim?    $1\frac{19}{20}$ miles

## Reteach
### 5-4 Regrouping to Subtract Mixed Numbers

You can use fraction strips to regroup to subtract mixed numbers.

To find $3\frac{1}{4} - 1\frac{3}{4}$, first model the first mixed number in the expression.

| 1 | 1 | 1 | $\frac{1}{4}$ |

There are not enough $\frac{1}{4}$ pieces to subtract, so you have to regroup.
Trade one one-whole strip for four $\frac{1}{4}$ pieces, because $\frac{4}{4} = 1$.

| 1 | 1 | $\frac{1}{4}$ $\frac{1}{4}$ $\frac{1}{4}$ $\frac{1}{4}$ | $\frac{1}{4}$ |

Now there are enough $\frac{1}{4}$ pieces to subtract. Take away $1\frac{3}{4}$.

| 1 | 1 | $\frac{1}{4}$ $\frac{1}{4}$ $\frac{1}{4}$ $\frac{1}{4}$ | $\frac{1}{4}$ |

The remaining pieces represent the difference. Write the difference in simplest form.

$3\frac{1}{4} - 1\frac{3}{4} = 1\frac{2}{4} = 1\frac{1}{2}$

Use fraction strips to find each difference. Write your answer in simplest form.

1. $3\frac{1}{4} - 2\frac{3}{4}$    $\frac{1}{2}$
2. $3\frac{1}{6} - 1\frac{5}{6}$    $1\frac{1}{3}$
3. $4\frac{3}{8} - 1\frac{7}{8}$    $2\frac{1}{2}$
4. $3\frac{1}{3} - 2\frac{2}{3}$    $\frac{2}{3}$
5. $5\frac{5}{12} - 2\frac{7}{12}$    $2\frac{5}{6}$
6. $3\frac{3}{10} - 1\frac{9}{10}$    $1\frac{2}{5}$
7. $5\frac{1}{8} - 1\frac{5}{8}$    $3\frac{1}{2}$
8. $4 - 1\frac{1}{3}$    $2\frac{2}{3}$
9. $3\frac{1}{8} - 1\frac{3}{8}$    $1\frac{3}{4}$
10. $2\frac{1}{8} - 1\frac{7}{8}$    $\frac{1}{4}$
11. $3 - 1\frac{1}{4}$    $1\frac{3}{4}$
12. $6\frac{3}{8} - 2\frac{5}{8}$    $3\frac{3}{4}$

Holt Mathematics

## Challenge
### 5-4 Popular First Names

What are the most popular first names in the United States?

Regroup fractions or mixed numbers to solve each problem below. Write your answers in simplest form. Then, in the box at the bottom of the page, write each problem's letter in the blanks above its solution. When you have solved all the problems, you will have found the answer to the question.

$8\frac{7}{12} - 7\frac{3}{4}$  $\quad \frac{5}{6}$  A

$9\frac{1}{8} - 8\frac{3}{4}$  $\quad \frac{3}{8}$  E

$10\frac{1}{3} - 9\frac{2}{3}$  $\quad \frac{2}{3}$  J

$6\frac{1}{2} - 5\frac{4}{5}$  $\quad \frac{7}{10}$  M

$5\frac{1}{5} - 4\frac{4}{5}$  $\quad \frac{2}{5}$  R

$7\frac{2}{9} - 6\frac{2}{3}$  $\quad \frac{5}{9}$  S

$12\frac{2}{5} - 11\frac{1}{2}$  $\quad \frac{9}{10}$  Y

#1 Name For American Men: J A M E S
$\quad \frac{2}{3} \quad \frac{5}{6} \quad \frac{7}{10} \quad \frac{3}{8} \quad \frac{5}{9}$

#1 Name For American Women: M A R Y
$\quad \frac{7}{10} \quad \frac{5}{6} \quad \frac{2}{5} \quad \frac{9}{10}$

## Problem Solving
### 5-4 Regrouping to Subtract Mixed Numbers

Write the correct answer in simplest form.

1. The average person in the United States eats $6\frac{13}{16}$ pounds of potato chips each year. The average person in Ireland eats $5\frac{15}{16}$ pounds. How much more potato chips do Americans eat a year than people in Ireland?
   $\frac{7}{8}$ pound more

2. The average person in the United States eats $270\frac{1}{16}$ pounds of meat each year. The average person in Australia eats $238\frac{1}{2}$ pounds. How much more meat do Americans eat a year than people in Australia?
   $31\frac{9}{16}$ pounds more

3. The average Americans eats $24\frac{1}{2}$ pounds of ice cream every year. The average person in Israel eats $15\frac{4}{5}$ pounds. How much more ice cream do Americans eat each year?
   $8\frac{7}{10}$ pounds more

4. People in Switzerland eat the most chocolate—26 pounds a year per person. Most Americans eat $12\frac{9}{16}$ pounds each year. How much more chocolate do the Swiss eat?
   $13\frac{7}{16}$ pounds more

5. The average person in the United States chews $1\frac{9}{16}$ pounds of gum each year. The average person in Japan chews $\frac{7}{8}$ pound. How much more gum do Americans chew?
   $\frac{11}{16}$ pound more

6. Norwegians eat the most frozen foods—$78\frac{1}{2}$ pounds per person each year. Most Americans eat $35\frac{15}{16}$ pounds. How much more frozen foods do people in Norway eat?
   $42\frac{9}{16}$ pounds more

Circle the letter of the correct answer.

7. Most people around the world eat $41\frac{7}{8}$ pounds of sugar each year. Most Americans eat $66\frac{3}{4}$ pounds. How much more sugar do Americans eat than the world's average?
   A $25\frac{7}{8}$ pounds more
   B $25\frac{1}{8}$ pounds more
   **C** $24\frac{7}{8}$ pounds more
   D $24\frac{1}{8}$ pounds more

8. The average person eats 208 pounds of vegetables and $125\frac{5}{8}$ pounds of fruit each year. How much more vegetables do most people eat than fruit?
   F $83\frac{5}{8}$ pounds more
   **G** $82\frac{3}{8}$ pounds more
   H $123\frac{5}{8}$ pounds more
   J $83\frac{3}{8}$ pounds more

## Reading Strategies
### 5-4 Compare and Contrast

When you subtract whole numbers, you often need to regroup a number before you can subtract.

$\quad 7\ 13$
$\quad 8\ \cancel{3}$
$-1\ 7$

In the above example, there were only 3 ones—not enough to subtract 7. A ten was regrouped as ten ones. The ten ones were added to the three ones to make 13 ones. Now there are enough ones to subtract 7.

You can compare regrouping fractions to regrouping whole numbers.

$\quad 3\frac{1}{8}$
$-1\frac{3}{8}$

Look at the fractions first. There aren't enough eighths to subtract $\frac{3}{8}$ from $\frac{1}{8}$. Regrouping fractions is different from regrouping whole numbers, because you regroup a whole number as a fraction. You can regroup 1 as a fraction with the same numerator and denominator.

$1 = \frac{2}{2} \quad 1 = \frac{5}{5} \quad 1 = \frac{8}{8}$

$2\frac{9}{8}$
$3\frac{1}{8}$
$-1\frac{3}{8}$

Take one from three and regroup as it $\frac{8}{8}$. Combine $\frac{8}{8}$ with $\frac{1}{8}$ to make $\frac{9}{8}$. Now there are enough eighths to subtract.

1. What is $\frac{9}{8} - \frac{3}{8}$? $\quad \frac{6}{8}$

2. What is the same about subtracting whole numbers and subtracting fractions?
   Numbers may have to be renamed.

3. What is different about subtracting whole numbers and subtracting fractions?
   When you subtract whole numbers, you rename whole numbers. With fractions you rename a whole number as a fraction.

## Puzzles, Twisters & Teasers
### 5-4 Subtraction Chains

Start with the first number in the chain. Subtract the next number, and the next, and the next. If, however, the next number to be subtracted is larger than your current answer, end the chain. Circle the last number you were able to subtract in that chain.

Example chain: | 7 | 2 | 4 | 3 |  7 − 2 = 5. 5 − 4 = 1. Stop now, because 3 is larger than your current answer. Circle 4, the last number you were able to subtract.

1. | $1\frac{2}{3}$ | $(1\frac{1}{3})$ | $\frac{2}{3}$ | $3\frac{1}{3}$ |
   |  O  |  H  |  N  |  J  |

2. | $3\frac{1}{7}$ | $1\frac{5}{7}$ | $\frac{5}{7}$ | $(\frac{4}{7})$ |
   |  U  |  I  |  M  |  E  |

3. | $5\frac{1}{5}$ | $3\frac{3}{5}$ | $(\frac{4}{5})$ | $1\frac{1}{5}$ |
   |  C  |  Y  |  T  |  H  |

4. | $4\frac{1}{8}$ | $(1\frac{5}{8})$ | $2\frac{5}{8}$ | $\frac{1}{8}$ |
   |  B  |  A  |  W  |  E  |

5. | $3\frac{3}{12}$ | $\frac{8}{12}$ | $\frac{9}{12}$ | $(\frac{7}{12})$ |
   |  V  |  F  |  R  |  S  |

Now you are ready to solve the riddle. Place the letters for the circled numbers in the numbered spaces and you will have your answer!

In a contest at a local restaurant, the restaurant owner hung two sirloins from the ceiling. Anyone who could jump up and get one won a free dinner. A customer came in, but when he was asked if he would like to try, he responded: "No thanks,

T H E   S T E A K S   A R E
3  1    2  5  3  2  4    5  4  2

T O O   H I G H
3        1     1

## Practice A
### 5-5 Solving Fraction Equations: Addition and Subtraction

Solve each equation. Write the solution in simplest form.

1. $k + 1\frac{1}{2} = 3$
   $k = 1\frac{1}{2}$

2. $m - 2\frac{1}{3} = 1\frac{1}{2}$
   $m = 3\frac{5}{6}$

3. $1\frac{1}{4} - \frac{2}{3} = p$
   $p = \frac{7}{12}$

4. $n + 3\frac{7}{8} = 5\frac{1}{8}$
   $n = 1\frac{1}{4}$

5. $3\frac{1}{3} = y - 1\frac{1}{6}$
   $y = 4\frac{1}{2}$

6. $2\frac{1}{5} + d = 3\frac{1}{2}$
   $d = 1\frac{3}{10}$

7. $2\frac{1}{7} + q = 4\frac{3}{14}$
   $q = 2\frac{1}{14}$

8. $z - 1\frac{2}{5} = 1\frac{7}{10}$
   $z = 3\frac{1}{10}$

9. $f + \frac{2}{3} = 1\frac{1}{9}$
   $f = \frac{4}{9}$

10. $b = 1\frac{5}{8} - \frac{3}{4}$
    $b = \frac{7}{8}$

11. $t + 1\frac{1}{5} = 3\frac{3}{10}$
    $t = 2\frac{1}{10}$

12. $3\frac{1}{2} + w = 5\frac{7}{12}$
    $w = 2\frac{1}{12}$

13. $c - 8\frac{1}{5} = 10\frac{3}{10}$
    $c = 18\frac{1}{2}$

14. $h + \frac{1}{3} = 2\frac{1}{6}$
    $h = 1\frac{5}{6}$

15. $1\frac{5}{9} = g - 3\frac{5}{18}$
    $g = 4\frac{5}{6}$

16. Joey beat Frank in the swim race by $2\frac{1}{10}$ minutes. Frank's time was $8\frac{3}{5}$ minutes. What was Joey's time in the race?
    $6\frac{1}{2}$ minutes

17. Sabrina bought 8 gallons of paint. After she painted her kitchen, she had $4\frac{1}{6}$ gallons left over. How much paint did Sabrina use in her kitchen?
    $3\frac{5}{6}$ gallons

## Practice B
### 5-5 Solving Fraction Equations: Addition and Subtraction

Solve each equation. Write the solution in simplest form. Check your answers.

1. $k + 3\frac{3}{4} = 5\frac{2}{3} - 1\frac{1}{3}$
   $k = \frac{7}{12}$

2. $a - 2\frac{2}{11} = 2\frac{5}{22} - 1\frac{2}{11}$
   $a = 3\frac{5}{22}$

3. $2\frac{2}{7} = n - 4\frac{2}{3} - 1\frac{1}{3}$
   $n = 8\frac{2}{7}$

4. $6\frac{1}{4} = z + 1\frac{5}{8}$
   $z = 4\frac{5}{8}$

5. $5\frac{1}{4} = x + \frac{7}{16}$
   $x = 4\frac{13}{16}$

6. $r + 6 = 9\frac{2}{5} - 2\frac{1}{2}$
   $r = \frac{9}{10}$

7. $11\frac{2}{5} = q - 4\frac{2}{7} + 2\frac{1}{7}$
   $q = 13\frac{19}{35}$

8. $4\frac{2}{5} - 2\frac{1}{2} = p + \frac{3}{10}$
   $p = 1\frac{3}{5}$

9. $\frac{3}{8} + \frac{1}{6} = c - 4\frac{5}{6}$
   $c = 5\frac{3}{8}$

10. $2\frac{1}{4} + c = 2\frac{1}{3} + 1\frac{1}{6}$
    $c = 1\frac{1}{4}$

11. A seamstress raised the hem on Helen's skirt by $1\frac{1}{3}$ inches. The skirt's original length was 16 inches. What is the new length?
    $14\frac{2}{3}$ inches

12. The bike trail is $5\frac{1}{4}$ miles long. Jessie has already cycled $2\frac{5}{8}$ miles of the trail. How much farther does she need to go to finish the trail?
    $2\frac{5}{8}$ miles

## Practice C
### 5-5 Solving Fraction Equations: Addition and Subtraction

Solve each equation. Write the solution in simplest form. Check your answers.

1. $3\frac{1}{5} - 1\frac{3}{10} = p + \frac{5}{6}$
   $p = 1\frac{1}{15}$

2. $17\frac{5}{6} + \frac{7}{10} = d - 2\frac{5}{12}$
   $d = 20\frac{19}{20}$

3. $34\frac{1}{6} = x + 6\frac{1}{4} + 12\frac{3}{8}$
   $x = 15\frac{13}{24}$

4. $a - 2\frac{3}{11} = 19\frac{1}{2} - 16\frac{1}{4}$
   $a = 5\frac{23}{44}$

5. $f - 4\frac{1}{10} + 15\frac{3}{5} = 29\frac{18}{25}$
   $f = 18\frac{11}{50}$

6. $\frac{7}{12} + \frac{3}{8} = c - 2\frac{5}{6}$
   $c = 3\frac{19}{24}$

7. $r + 11\frac{3}{5} = 20\frac{1}{5} - 3\frac{1}{2}$
   $r = 5\frac{1}{10}$

8. $s + 30\frac{11}{15} = 40\frac{1}{3} - 2\frac{1}{2}$
   $s = 7\frac{1}{10}$

9. Carol wants each of the curtains she makes to be the same length. She started with two pieces of cloth measuring $6\frac{1}{3}$ feet and $7\frac{3}{4}$ feet. She cut $1\frac{5}{8}$ feet off the $6\frac{1}{3}$-foot piece. How much should she cut from the second piece?
   $3\frac{1}{24}$ feet

10. Last year it rained $42\frac{1}{6}$ inches in Portland, Maine. It rained $14\frac{5}{8}$ inches in the spring, and $11\frac{1}{24}$ inches in the summer. The city received the same amounts of rain in the fall as in winter. How much did it rain in fall?
    $8\frac{1}{4}$ inches

## Reteach
### 5-5 Solving Fraction Equations: Addition and Subtraction

You can write related facts using addition and subtraction.
$3 + 4 = 7$     $7 - 4 = 3$
You can use related facts to solve equations.

A. $x + 2\frac{1}{2} = 4$
   Think: $4 - 2\frac{1}{2} = x$
   $x = 4 - 2\frac{1}{2}$
   $x = 3\frac{2}{2} - 2\frac{1}{2}$   Regroup 4 as $3\frac{2}{2}$.
   $x = 1\frac{1}{2}$

B. $x - 4\frac{1}{3} = 3\frac{1}{2}$
   Think: $3\frac{1}{2} + 4\frac{1}{3} = x$
   $x = 3\frac{1}{2} + 4\frac{1}{3}$
   $x = \frac{7}{2} + \frac{13}{3}$     Write the mixed numbers as improper fractions.
   $x = \frac{21}{6} + \frac{26}{6}$    Write the fractions using a common denominator.
   $x = \frac{47}{6}$
   $x = 7\frac{5}{6}$                   Write the sum as a mixed number.

Use related facts to solve each equation.

1. $x + 3\frac{1}{3} = 7$
   $x = 7 - 3\frac{1}{3}$
   $x = 6\frac{3}{3} - 3\frac{1}{3}$
   $x = 3\frac{2}{3}$

2. $x - 2\frac{1}{4} = 4\frac{1}{2}$
   $x = 4\frac{1}{2} + 2\frac{1}{4}$
   $x = \frac{9}{2} + \frac{9}{4}$
   $x = \frac{18}{4} + \frac{9}{4}$
   $x = 6\frac{3}{4}$

3. $x + \frac{3}{8} = 5\frac{1}{4}$
   $x = 5\frac{1}{4} - \frac{3}{8}$
   $x = \frac{21}{4} - \frac{3}{8}$
   $x = \frac{42}{8} - \frac{3}{8}$
   $x = 4\frac{7}{8}$

4. $x - \frac{5}{12} = 2\frac{1}{2}$
   $x = 2\frac{1}{2} + \frac{5}{12}$
   $x = \frac{5}{2} + \frac{5}{12}$
   $x = \frac{30}{12} + \frac{5}{12}$
   $x = 2\frac{11}{12}$

5. $x - 1\frac{3}{4} = 7\frac{1}{2}$
   $x = 9\frac{1}{4}$

6. $x - 3\frac{2}{3} = 1\frac{1}{3}$
   $x = 5$

7. $x + 3\frac{1}{2} = 6\frac{1}{4}$
   $x = 2\frac{3}{4}$

8. $x - 2\frac{2}{5} = 1\frac{3}{10}$
   $x = 3\frac{7}{10}$

## Challenge
### 5-5 You Read My Mind!

Here's a trick you can use to amaze your friends and family. Start by asking your friends to think of a number—any number. Pretend you are reading their minds while you write the number 6 on a piece of paper. (Don't show it to them.) Then use the steps below to tell them what to do. The fraction $\frac{2}{5}$ is used as an example choice, but the trick works for any chosen fraction, mixed number, decimal, or whole number.

| Step | Example |
| --- | --- |
| 1. Double your number. | $\frac{2}{5} + \frac{2}{5} = \frac{4}{5}$ |
| 2. Add 12 to your sum. | $\frac{4}{5} + 12 = 12\frac{4}{5}$ |
| 3. Divide your new sum by 2. | $12\frac{4}{5} \div 2 = 6\frac{2}{5}$ |
| 4. Subtract your chosen number from that quotient. | $6\frac{2}{5} - \frac{2}{5} = 6$ |

Now amaze your friends by showing that you wrote the same number they ended with. No matter what number is chosen, this trick always ends in 6. Equations explain why it works—but don't tell your friends this part.

Let $x$ = the chosen number.

STEP 1 → $2x$
STEP 2 → $2x + 12$
STEP 3 → $(2x + 12) \div 2 = x + 6$
STEP 4 → $x + 6 - x = 6$

Before you try the trick, practice it on the fractions below. Use the equations for each step and show all your work.

1. Chosen Number: $\frac{7}{9}$

STEP 1: $2 \cdot \frac{7}{9} = \frac{14}{9} = 1\frac{5}{9}$
STEP 2: $1\frac{5}{9} + 12 = 13\frac{5}{9}$
STEP 3: $13\frac{5}{9} \div 2 = \frac{122}{18} = 6\frac{14}{18} = 6\frac{7}{9}$
STEP 4: $6\frac{7}{9} - \frac{7}{9} = 6$

2. Chosen Number: $3\frac{1}{4}$

STEP 1: $2 \cdot 3\frac{1}{4} = \frac{26}{4} = 6\frac{2}{4} = 6\frac{1}{2}$
STEP 2: $6\frac{1}{2} + 12 = 18\frac{1}{2}$
STEP 3: $18\frac{1}{2} \div 2 = \frac{37}{4} = 9\frac{1}{4}$
STEP 4: $9\frac{1}{4} - 3\frac{1}{4} = 6$

## Problem Solving
### 5-5 Solving Fraction Equations: Addition and Subtraction

Write the correct answer in simplest form.

1. It usually takes Brian $1\frac{1}{2}$ hours to get to work from the time he gets out of bed. His drive to the office takes $\frac{3}{4}$ hour. How much time does he spend getting ready for work?

   $\frac{3}{4}$ of an hour

2. Before she went to the hairdresser, Sheila's hair was $7\frac{1}{4}$ inches long. When she left the salon, it was $5\frac{1}{2}$ inches long. How much of her hair did Sheila get cut off?

   $1\frac{3}{4}$ inches

3. One lap around the gym is $\frac{1}{3}$ mile long. Kim has already run 5 times around. If she wants to run 2 miles total, how much farther does she have to go?

   $\frac{1}{3}$ mile more

4. Darius timed his speech at $5\frac{1}{6}$ minutes. His time limit for the speech is $4\frac{1}{2}$ minutes. How much does he need to cut from his speech?

   $\frac{2}{3}$ minute

Circle the letter of the correct answer.

5. Mei and Alex bought the same amount of food at the deli. Mei bought $1\frac{1}{4}$ pounds of turkey and $1\frac{1}{3}$ pounds of cheese. Alex bought $1\frac{1}{2}$ pounds of turkey. How much cheese did Alex buy?

   **A** $1\frac{1}{12}$ pounds    C $1\frac{1}{4}$ pounds
   B $1\frac{1}{6}$ pounds    D $4\frac{1}{12}$ pounds

6. When Lynn got her dog, Max, he weighed $10\frac{1}{2}$ pounds. During the next 6 months, he gained $8\frac{4}{5}$ pounds. At his one-year check-up he had gained another $4\frac{1}{3}$ pounds. How much did Max weigh when he was 1 year old?

   F $22\frac{19}{30}$ pounds    H $23\frac{29}{30}$ pounds
   **G** $23\frac{19}{30}$ pounds    J $23\frac{49}{50}$ pounds

7. Charlie picked up 2 planks of wood at the hardware store. One is $6\frac{1}{4}$ feet long and the other is $5\frac{5}{8}$ feet long. How much should he cut from the first plank to make them the same length?

   **A** $\frac{5}{8}$ foot    C $1\frac{3}{8}$ feet
   B $\frac{1}{2}$ foot    D $1\frac{5}{8}$ feet

8. Carmen used $3\frac{3}{4}$ cups of flour to make a cake. She had $\frac{1}{2}$ cup of flour left over. Which equation can you use to find how much flour she had before baking the cake?

   F $x + \frac{1}{2} = 3\frac{3}{4}$    H $3\frac{3}{4} - \frac{1}{2} = x$
   **G** $x - 3\frac{3}{4} = \frac{1}{2}$    J $3\frac{3}{4} - x = \frac{1}{2}$

## Reading Strategies
### 5-5 Summarize

The following steps are used to solve addition and subtraction equations with fractions.

$2\frac{1}{3} + m = 5$
$2\frac{1}{3} - 2\frac{1}{3} + m = 5 - 2\frac{1}{3}$ ← Step 1: Subtract $2\frac{1}{3}$ from both sides of the equation.
$m = 5 - 2\frac{1}{3}$
$m = 4\frac{3}{3} - 2\frac{1}{3}$ ← Step 2: Regroup 5 as $4\frac{3}{3}$
$m = 2\frac{2}{3}$ ← Step 3: Subtract fractions. Subtract whole numbers.

Answer each question.

1. What is the first step in the example above?

   Subtract $2\frac{1}{3}$ from both sides of the equation.

2. Why was $2\frac{1}{3}$ subtracted from both sides of the equation?

   To get $m$ by itself.

3. What is the second step in the example above?

   Rename 5 as $4\frac{3}{3}$.

Use this equation to answer the following questions.

$x - 3\frac{2}{3} = 2\frac{2}{3}$

4. What is the first step to solve the equation?

   Add $3\frac{2}{3}$ to both sides of the equation.

5. What is the next step to solve the equation?

   Add fractions and whole numbers.

6. Write how you solve equations that involve fractions.

   Possible answer: Get the variable on one side of the equation, rename if needed, add or subtract fractions, and add or subtract whole numbers.

## Puzzles, Twisters & Teasers
### 5-5 Leap to Success

Solve the equations below and record the answers. Match the answer to the letter of the variable in the equation the answer goes with.

$R - \frac{4}{5} = \frac{1}{10}$    $R = \frac{9}{10}$

$2\frac{2}{3} + O = 3\frac{1}{8}$    $O = \frac{11}{24}$

$C + 6\frac{7}{8} = 10\frac{1}{5}$    $C = 3\frac{13}{40}$

$I - \frac{1}{11} = \frac{9}{22}$    $I = \frac{1}{2}$

$N - 2\frac{7}{9} = 1\frac{1}{2}$    $N = 4\frac{5}{18}$

$8\frac{3}{5} = A + 4\frac{6}{7}$    $A = 3\frac{26}{35}$

$5\frac{1}{4} + K = 6\frac{5}{12}$    $K = 1\frac{1}{6}$

$G + \frac{4}{5} = 2\frac{2}{21}$    $G = 1\frac{31}{105}$

Why were frogs put on the endangered species list?

Because they are always   C   R   O   A   K   I   N   G !
$3\frac{13}{40}$   $\frac{9}{10}$   $\frac{11}{24}$   $3\frac{26}{35}$   $1\frac{1}{6}$   $\frac{1}{2}$   $4\frac{5}{18}$   $1\frac{31}{105}$

## Practice A
### 5-6 Multiplying Fractions Using Repeated Addition

Multiply. Write each answer in simplest form.

1. $1 \cdot \frac{1}{3}$ = $\frac{1}{3}$
2. $3 \cdot \frac{1}{8}$ = $\frac{3}{8}$
3. $7 \cdot \frac{1}{9}$ = $\frac{7}{9}$
4. $3 \cdot \frac{1}{4}$ = $\frac{3}{4}$
5. $4 \cdot \frac{2}{10}$ = $\frac{4}{5}$
6. $3 \cdot \frac{1}{6}$ = $\frac{1}{2}$
7. $2 \cdot \frac{2}{5}$ = $\frac{4}{5}$
8. $10 \cdot \frac{1}{2}$ = $5$
9. $5 \cdot \frac{1}{8}$ = $\frac{5}{8}$
10. $4 \cdot \frac{1}{6}$ = $\frac{2}{3}$
11. $5 \cdot \frac{1}{8}$ = $\frac{5}{8}$
12. $3 \cdot \frac{2}{6}$ = $1$
13. $7 \cdot \frac{1}{11}$ = $\frac{7}{11}$
14. $3 \cdot \frac{1}{9}$ = $\frac{1}{3}$
15. $5 \cdot \frac{1}{15}$ = $\frac{1}{3}$

Evaluate $2x$ for each value of $x$. Write the answer in simplest form.

16. $x = \frac{1}{4}$ = $\frac{1}{2}$
17. $x = \frac{1}{3}$ = $\frac{2}{3}$
18. $x = \frac{1}{2}$ = $1$
19. $x = \frac{1}{6}$ = $\frac{1}{3}$
20. $x = \frac{1}{7}$ = $\frac{2}{7}$
21. $x = \frac{1}{8}$ = $\frac{1}{4}$
22. $x = \frac{2}{3}$ = $1\frac{1}{3}$
23. $x = \frac{3}{4}$ = $1\frac{1}{2}$

24. Richie is making 3 quarts of fruit punch for his friends. He must add $\frac{1}{2}$ cup sugar to make each quart of punch. How much sugar will he add?
$1\frac{1}{2}$ cups

25. Mrs. Flynn has 20 students in her class. One-fourth of her students purchased lunch tokens. How many of her students purchased tokens?
5 students

---

## Practice B
### 5-6 Multiplying Fractions Using Repeated Addition

Multiply. Write each answer in simplest form.

1. $5 \cdot \frac{1}{10}$ = $\frac{1}{2}$
2. $6 \cdot \frac{1}{18}$ = $\frac{1}{3}$
3. $4 \cdot \frac{1}{14}$ = $\frac{2}{7}$
4. $3 \cdot \frac{1}{12}$ = $\frac{1}{4}$
5. $2 \cdot \frac{1}{8}$ = $\frac{1}{4}$
6. $6 \cdot \frac{1}{10}$ = $\frac{3}{5}$
7. $3 \cdot \frac{1}{6}$ = $\frac{1}{2}$
8. $3 \cdot \frac{5}{12}$ = $1\frac{1}{4}$
9. $3 \cdot \frac{2}{7}$ = $\frac{6}{7}$
10. $2 \cdot \frac{3}{8}$ = $\frac{3}{4}$
11. $10 \cdot \frac{3}{15}$ = $2$
12. $8 \cdot \frac{2}{14}$ = $1\frac{1}{7}$
13. $5 \cdot \frac{2}{10}$ = $1$
14. $4 \cdot \frac{4}{12}$ = $1\frac{1}{3}$
15. $2 \cdot \frac{13}{20}$ = $1\frac{3}{10}$

Evaluate $6x$ for each value of $x$. Write the answer in simplest form.

16. $x = \frac{2}{3}$ = $4$
17. $x = \frac{2}{8}$ = $1\frac{1}{2}$
18. $x = \frac{1}{4}$ = $1\frac{1}{2}$
19. $x = \frac{2}{6}$ = $2$
20. $x = \frac{2}{7}$ = $1\frac{5}{7}$
21. $x = \frac{2}{5}$ = $2\frac{2}{5}$
22. $x = \frac{3}{11}$ = $1\frac{7}{11}$
23. $x = \frac{5}{12}$ = $2\frac{1}{2}$

24. Thomas spends 60 minutes exercising. For $\frac{1}{4}$ of that time, he jumps rope. How many minutes does he spend jumping rope?
15 minutes

25. Kylie made a 4-ounce milk shake. Two-thirds of the milk shake was ice cream. How many ounces of ice cream did Kylie use in the shake?
$2\frac{2}{3}$ ounces

---

## Practice C
### 5-6 Multiplying Fractions Using Repeated Addition

Multiply. Write each answer in simplest form.

1. $3 \cdot \frac{4}{17}$ = $\frac{12}{17}$
2. $2 \cdot \frac{6}{10}$ = $1\frac{1}{5}$
3. $4 \cdot \frac{3}{4}$ = $3$
4. $5 \cdot \frac{6}{15}$ = $2$
5. $3 \cdot \frac{8}{9}$ = $2\frac{2}{3}$
6. $6 \cdot \frac{3}{14}$ = $1\frac{2}{7}$
7. $12 \cdot \frac{3}{42}$ = $\frac{6}{7}$
8. $6 \cdot \frac{4}{27}$ = $\frac{8}{9}$
9. $2 \cdot \frac{16}{20}$ = $1\frac{3}{5}$

Evaluate $9x$ for each value of $x$. Write the answer in simplest form.

10. $x = \frac{2}{9}$ = $2$
11. $x = \frac{3}{36}$ = $\frac{3}{4}$
12. $x = \frac{9}{18}$ = $4\frac{1}{2}$
13. $x = \frac{4}{7}$ = $5\frac{1}{7}$
14. $x = \frac{5}{12}$ = $3\frac{3}{4}$
15. $x = \frac{4}{81}$ = $\frac{4}{9}$
16. $x = \frac{16}{18}$ = $8$
17. $x = \frac{27}{50}$ = $4\frac{43}{50}$

Evaluate each expression. Write each answer in simplest form.

18. $9c$ for $c = \frac{2}{3}$ = $6$
19. $13d$ for $d = \frac{2}{9}$ = $2\frac{8}{9}$
20. $7n$ for $n = \frac{1}{7}$ = $1$

Compare. Write <, >, or =.

21. $3 \cdot \frac{1}{2}$ > $\frac{4}{7}$
22. $\frac{3}{5}$ = $3 \cdot \frac{1}{5}$
23. $12 \cdot \frac{3}{4}$ < $10$

24. Clair's paycheck this week was $568.00. She put $\frac{1}{4}$ of that amount in her savings account. Then she spent $\frac{1}{2}$ of what was left on rent and $42.60 on groceries. How much money does she have left?
$170.40

25. A television news program questioned 270 people to see if they voted in the election. Of those questioned, $\frac{2}{15}$ did not vote in the election. How many of those questioned did vote?
234 people voted

---

## Reteach
### 5-6 Multiplying Fractions Using Repeated Addition

You can use fraction strips to multiply fractions by whole numbers.
To find $3 \cdot \frac{2}{3}$, first think about the expression in words.
$3 \cdot \frac{2}{3}$ means "3 groups of $\frac{2}{3}$."
Then model the expression.

The total number of $\frac{1}{3}$ fraction pieces is 6.
So, $3 \cdot \frac{2}{3} = \frac{2}{3} + \frac{2}{3} + \frac{2}{3} =$
$\frac{6}{3} = 2$ in simplest form.

Use fraction strips to find each product.

1. $4 \cdot \frac{1}{8}$ = $\frac{1}{2}$
2. $2 \cdot \frac{2}{5}$ = $\frac{4}{5}$
3. $6 \cdot \frac{1}{8}$ = $\frac{3}{4}$
4. $8 \cdot \frac{1}{4}$ = $2$

You can also use counters to multiply fractions by whole numbers.
To find $\frac{1}{2} \cdot 12$, first think about the expression in words.
$\frac{1}{2} \cdot 12 = \frac{12}{2}$, which means "12 divided into 2 equal groups."
Then model the expression.

The number of counters in each group is the product.
$\frac{1}{2} \cdot 12 = 6$.

Use counters to find each product.

5. $\frac{1}{3} \cdot 15$ = $5$
6. $\frac{1}{8} \cdot 24$ = $3$
7. $\frac{1}{4} \cdot 16$ = $4$
8. $\frac{1}{12} \cdot 24$ = $2$

---

Holt Mathematics

## Challenge
### 5-6 Slowpoke Race

The animals shown below are some of the slowest creatures on Earth. Use their given average speeds to find how far they will travel in the times marked along their racetracks.

Which of these slowpokes traveled the farthest? __three-toed sloth__

Three-toed sloth
Speed: $\frac{3}{5}$ mi/h

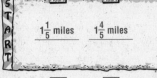
2 Hours: $1\frac{1}{5}$ miles  3 Hours: $1\frac{4}{5}$ miles  5 Hours: 3 miles

Earthworm
Speed: $\frac{1}{10}$ mi/h

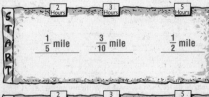
2 Hours: $\frac{1}{5}$ mile  3 Hours: $\frac{3}{10}$ mile  5 Hours: $\frac{1}{2}$ mile

Tortoise
Speed: $\frac{1}{5}$ mi/h

2 Hours: $\frac{2}{5}$ mile  3 Hours: $\frac{3}{5}$ mile  5 Hours: 1 mile

Snail
Speed: $\frac{3}{10}$ mi/h

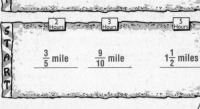
2 Hours: $\frac{3}{5}$ mile  3 Hours: $\frac{9}{10}$ mile  5 Hours: $1\frac{1}{2}$ miles

---

## Problem Solving
### 5-6 Multiplying Fractions Using Repeated Addition

Write the answers in simplest form.

1. Did you know that some people have more bones than the rest of the population? About $\frac{1}{20}$ of all people have an extra rib bone. In a crowd of 60 people, about how many people are likely have an extra rib bone?

   __3 people__

2. The Appalachian National Scenic Trail is the longest marked walking path in the United States. It extends through 14 states for about 2,000 miles. Last year, Carla hiked $\frac{1}{5}$ of the trail. How many miles of the trail did she hike?

   __400 miles__

3. Human fingernails can grow up to $\frac{1}{10}$ of a millimeter each day. How much can fingernails grow in one week?

   $\frac{7}{10}$ millimeter

4. Most people dream about $\frac{1}{4}$ of the time they sleep. How long will you probably dream tonight if you sleep for 8 hours?

   __2 hours__

Circle the letter of the correct answer.

5. Today, the United States flag has 50 stars—one for each state. The first official U.S. flag was approved in 1795. It had $\frac{3}{10}$ as many stars as today's flag. How many stars were on the first official U.S. flag?
   - A  5 stars
   - B  10 stars
   - **C  15 stars**
   - D  35 stars

6. The Statue of Liberty is about 305 feet tall from the ground to the tip of her torch. The statue's pedestal makes up about $\frac{1}{2}$ of its height. About how tall is the pedestal of the Statue of Liberty?
   - F  610 feet
   - **G  152 1/2 feet**
   - H  150 1/2 feet
   - J  102 1/2 feet

7. The Caldwells own a 60-acre farm. They planted $\frac{3}{5}$ of the land with corn. How many acres of corn did they plant?
   - A  12 acres
   - **B  36 acres**
   - C  20 acres
   - D  18 acres

8. Objects on Uranus weigh about $\frac{4}{5}$ of their weight on Earth. If a dog weighs 40 pounds on Earth, how much would it weigh on Uranus?
   - **F  32 pounds**
   - G  10 pounds
   - H  8 pounds
   - J  30 pounds

---

## Reading Strategies
### 5-6 Relate Words and Symbols

Repeated addition is a way to represent multiplication of fractions.

$\frac{1}{8} + \frac{1}{8} + \frac{1}{8} = \frac{3}{8}$ → Repeated addition
three times one-eighth = three-eighths → Words
$3 \cdot \frac{1}{8} = \frac{3}{8}$ → Symbols

Answer the following questions.

1. What is $\frac{2}{8} \cdot 2$? __$\frac{4}{8}$__

2. What is three-eighths times two? __six-eighths__

3. What is $\frac{1}{8} \cdot 4$? __$\frac{4}{8}$__

4. Write $\frac{1}{8} + \frac{1}{8} + \frac{1}{8} + \frac{1}{8}$ as a multiplication problem.  $4 \cdot \frac{1}{8} = \frac{4}{8}$

Use the rectangle to answer each question.

5. What is two-tenths times two? __four-tenths__

6. What is $\frac{1}{10} \cdot 4$? __$\frac{4}{10}$__

7. What is four-tenths times two? __eight-tenths__

8. Write $\frac{1}{10} + \frac{1}{10} + \frac{1}{10} + \frac{1}{10}$ as a multiplication problem in words. __four times one-tenth = four tenths__

---

## Puzzles, Twisters & Teasers
### 5-6 Run-Away Computers

Fill in the blanks to complete each statement. Match your answers to the letters to solve the riddle.

1. $\frac{1}{4} \times 16$  __4__  T
2. $\frac{1}{6}$ of 18  __3__  W
3. $\frac{1}{5}$ of 50  __10__  E
4. $\frac{7}{11} \times 22$  __14__  R
5. $\frac{4}{5}$ of 15  __12__  N
6. $\frac{1}{2} \times 34$  __17__  E
7. $\frac{1}{9} \times 45$  __5__  V
8. $\frac{3}{4} \times 28$  __21__  N
9. $\frac{3}{5} \times 40$  __24__  I
10. $\frac{1}{2}$ of 100  __50__  T

How do you catch a runaway computer?

With an  I  N  T  E  R  N  E  T
         24 21  4  10 14 12 17 50

## Practice A
### 5-7 Multiplying Fractions

Multiply. Write each answer in simplest form.

1. $\frac{1}{2} \cdot \frac{1}{7}$ = $\frac{1}{14}$
2. $\frac{1}{4} \cdot \frac{1}{4}$ = $\frac{1}{16}$
3. $\frac{1}{5} \cdot \frac{1}{3}$ = $\frac{1}{15}$
4. $\frac{2}{3} \cdot \frac{1}{3}$ = $\frac{2}{9}$
5. $\frac{2}{3} \cdot \frac{2}{7}$ = $\frac{4}{21}$
6. $\frac{1}{4} \cdot \frac{1}{5}$ = $\frac{1}{20}$
7. $\frac{1}{3} \cdot \frac{2}{5}$ = $\frac{2}{15}$
8. $\frac{1}{4} \cdot \frac{2}{3}$ = $\frac{1}{6}$
9. $\frac{1}{3} \cdot \frac{1}{3}$ = $\frac{1}{9}$

Evaluate the expression $x \cdot \frac{1}{2}$ for each value of $x$. Write the answer in simplest form.

10. $x = \frac{1}{2}$ → $\frac{1}{4}$
11. $x = \frac{1}{3}$ → $\frac{1}{6}$
12. $x = \frac{1}{4}$ → $\frac{1}{8}$
13. $x = \frac{1}{5}$ → $\frac{1}{10}$
14. $x = \frac{2}{3}$ → $\frac{1}{3}$
15. $x = \frac{3}{4}$ → $\frac{3}{8}$

16. In Mr. Sanders's class, $\frac{1}{3}$ of the students are girls. About $\frac{1}{4}$ of the girls want to join the chorus. What fraction of all the students in Mr. Sanders's class want to join the chorus?
$\frac{1}{12}$ of the students

17. A recipe for trail mix calls for $\frac{3}{4}$ pound of peanuts. Luiza only wants to make half of the recipe's servings. How many pounds of peanuts should she use?
$\frac{3}{8}$ pound

## Practice B
### 5-7 Multiplying Fractions

Multiply. Write each answer in simplest form.

1. $\frac{1}{2} \cdot \frac{2}{5}$ = $\frac{1}{5}$
2. $\frac{1}{3} \cdot \frac{7}{8}$ = $\frac{7}{24}$
3. $\frac{2}{3} \cdot \frac{4}{6}$ = $\frac{4}{9}$
4. $\frac{1}{4} \cdot \frac{10}{11}$ = $\frac{5}{22}$
5. $\frac{3}{5} \cdot \frac{2}{3}$ = $\frac{2}{5}$
6. $\frac{8}{9} \cdot \frac{3}{4}$ = $\frac{2}{3}$
7. $\frac{3}{8} \cdot \frac{4}{5}$ = $\frac{3}{10}$
8. $\frac{2}{7} \cdot \frac{3}{4}$ = $\frac{3}{14}$
9. $\frac{1}{6} \cdot \frac{2}{3}$ = $\frac{1}{9}$

Evaluate the expression $x \cdot \frac{1}{5}$ for each value of $x$. Write the answer in simplest form.

10. $x = \frac{3}{7}$ → $\frac{3}{35}$
11. $x = \frac{5}{6}$ → $\frac{1}{6}$
12. $x = \frac{2}{3}$ → $\frac{2}{15}$
13. $x = \frac{10}{11}$ → $\frac{2}{11}$
14. $x = \frac{5}{8}$ → $\frac{1}{8}$
15. $x = \frac{4}{5}$ → $\frac{4}{25}$

16. A cookie recipe calls for $\frac{2}{3}$ cup of brown sugar. Sarah is making $\frac{1}{4}$ of the recipe. How much brown sugar will she need?
$\frac{1}{6}$ cup

17. Nancy spent $\frac{7}{8}$ hour working out at the gym. She spent $\frac{5}{7}$ of that time lifting weights. What fraction of an hour did she spend lifting weights?
$\frac{5}{8}$ hour

## Practice C
### 5-7 Multiplying Fractions

Multiply. Write each answer in simplest form.

1. $\frac{3}{8} \cdot \frac{4}{5}$ = $\frac{3}{10}$
2. $\frac{5}{8} \cdot \frac{3}{9}$ = $\frac{5}{24}$
3. $\frac{6}{7} \cdot \frac{5}{6}$ = $\frac{5}{7}$
4. $\frac{8}{9} \cdot \frac{9}{11}$ = $\frac{8}{11}$
5. $\frac{5}{12} \cdot \frac{6}{7}$ = $\frac{5}{14}$
6. $\frac{7}{9} \cdot \frac{3}{8}$ = $\frac{7}{24}$
7. $\frac{14}{15} \cdot \frac{5}{7}$ = $\frac{2}{3}$
8. $\frac{7}{8} \cdot \frac{2}{9}$ = $\frac{7}{36}$
9. $\frac{4}{5} \cdot \frac{7}{9} \cdot \frac{1}{7}$ = $\frac{4}{45}$

Evaluate the expression  for each value of $x$. Write the answer in simplest form.

10. $x = \frac{4}{5}$ → $\frac{8}{35}$
11. $x = \frac{7}{8}$ → $\frac{1}{4}$
12. $x = \frac{7}{11}$ → $\frac{2}{11}$
13. $x = \frac{11}{10}$ → $\frac{11}{35}$
14. $x = \frac{8}{9}$ → $\frac{16}{63}$
15. $x = \frac{21}{30}$ → $\frac{1}{5}$

Compare. Write <, >, or =.

16. $\frac{5}{6} \cdot \frac{3}{4}$ < $\frac{7}{8} \cdot \frac{4}{5}$
17. $\frac{2}{3} \cdot \frac{6}{7}$ > $\frac{9}{10} \cdot \frac{1}{3}$
18. $\frac{10}{12} \cdot \frac{5}{6}$ > $\frac{5}{9} \cdot \frac{1}{4}$
19. $\frac{7}{9} \cdot \frac{3}{4}$ = $\frac{7}{6} \cdot \frac{1}{2}$
20. $\frac{9}{11} \cdot \frac{1}{2}$ > $\frac{1}{6} \cdot \frac{3}{11}$
21. $\frac{2}{3} \cdot \frac{9}{10}$ < $\frac{4}{5} \cdot \frac{7}{8}$

22. Cara bought 1 yard of velvet at the fabric store. She used $\frac{5}{9}$ yard to make a purse. Then she used $\frac{1}{2}$ of the leftover velvet to make a hair band. How much of the velvet did she use to make the hair band?
$\frac{2}{9}$ yard

23. A square-shaped park measures $\frac{3}{5}$ mile long on each side. What is the area of the park?
$\frac{9}{25}$ square mile

## Reteach
### 5-7 Multiplying Fractions

To multiply fractions, multiply the numerators and multiply the denominators.

When multiplying fractions, you can sometimes divide by the GCF to make the problem simpler.

You can divide by the GCF even if the numerator and denominator of the same fraction have a common factor.

$\frac{1}{2} \cdot \frac{2}{3}$

$\frac{1}{\cancel{2}} \cdot \frac{\cancel{2}}{3}$

The problem is now $\frac{1}{1} \cdot \frac{1}{3}$.

$\frac{1 \cdot 1}{1 \cdot 3} = \frac{1}{3}$

So, $\frac{1}{2} \cdot \frac{2}{3} = \frac{1}{3}$.

Is it possible to simplify before you multiply? If so, what is the GCF?

1. $\frac{1}{4} \cdot \frac{1}{2}$ — no
2. $\frac{1}{6} \cdot \frac{3}{4}$ — yes; 3
3. $\frac{1}{8} \cdot \frac{2}{3}$ — yes; 2
4. $\frac{1}{3} \cdot \frac{2}{5}$ — no

Multiply.

5. $\frac{1}{6} \cdot \frac{3}{5}$ = $\frac{1}{10}$
6. $\frac{1}{4} \cdot \frac{1}{3}$ = $\frac{1}{12}$
7. $\frac{7}{8} \cdot \frac{4}{5}$ = $\frac{7}{10}$
8. $\frac{1}{6} \cdot \frac{2}{3}$ = $\frac{1}{9}$
9. $\frac{1}{5} \cdot \frac{1}{2}$ = $\frac{1}{10}$
10. $\frac{3}{5} \cdot \frac{1}{4}$ = $\frac{3}{20}$
11. $\frac{3}{7} \cdot \frac{1}{9}$ = $\frac{1}{21}$
12. $\frac{3}{4} \cdot \frac{1}{2}$ = $\frac{3}{8}$
13. $\frac{1}{3} \cdot \frac{6}{7}$ = $\frac{2}{7}$
14. $\frac{1}{4} \cdot \frac{2}{3}$ = $\frac{1}{6}$
15. $\frac{3}{4} \cdot \frac{1}{3}$ = $\frac{1}{4}$
16. $\frac{1}{4} \cdot \frac{1}{8}$ = $\frac{1}{32}$

## Challenge
### 5-7 Fractions of Flowers

For each flower below, shade the two petals whose fractions have a product equal to the fraction written in the center of that flower.

1.
2.
3.
4.
5.
6.

---

## Problem Solving
### 5-7 Multiplying Fractions

Use the circle graph to answer the questions. Write each answer in simplest form.

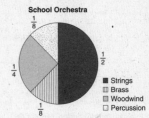

1. Of the students playing stringed instruments, $\frac{3}{4}$ play the violin. What fraction of the whole orchestra is violin players?

   $\frac{3}{8}$ of the orchestra

2. Of the students playing woodwind instruments, $\frac{1}{2}$ play the clarinet. What fraction of the whole orchestra is clarinet players?

   $\frac{1}{8}$ of the orchestra

Circle the letter of the correct answer.

3. Two-thirds of the students who play a percussion instrument are boys. What fraction of the musicians in the orchestra is boys who play percussion? girls who play percussion?
   - A $\frac{1}{24}$ of the orchestra
   - **B** $\frac{1}{12}$ of the orchestra
   - C $\frac{1}{4}$ of the orchestra
   - D $\frac{2}{3}$ of the orchestra

4. The brass section is evenly divided into horns, trumpets, trombones, and tubas. What fraction of the whole orchestra do players of each of those brass instruments make up?
   - F $\frac{1}{32}$ of the orchestra
   - G $\frac{1}{8}$ of the orchestra
   - H $\frac{1}{4}$ of the orchestra
   - J $\frac{1}{2}$ of the orchestra

5. There are 40 students in the orchestra. How many students play either percussion or brass instruments?
   - A 5 students
   - **B** 10 students
   - C 8 students
   - D 16 students

6. If 2 more violinists join the orchestra, what fraction of all the musicians would play a stringed instrument?
   - F $\frac{11}{21}$
   - G $\frac{11}{20}$
   - H $\frac{1}{20}$
   - J $\frac{1}{26}$

---

## Reading Strategies
### 5-7 Use Graphic Aids

The circle below is divided into two equal parts. Each part is equal to one-half.

If one-half of the circle is split in half, it looks like this.

$\frac{1}{2}$ of $\frac{1}{2}$ is $\frac{1}{4}$

$\frac{1}{2} \cdot \frac{1}{2} = \frac{1}{4}$

The drawing shows a rectangle divided into thirds.

1. If you divide $\frac{1}{3}$ of the rectangle in half, what fractional part will that be? $\frac{1}{6}$

2. One-half of $\frac{1}{3}$ = $\frac{1}{6}$

3. $\frac{1}{2} \cdot \frac{1}{3}$ = $\frac{1}{6}$

To multiply fractions:

$\frac{1}{2} \cdot \frac{1}{4}$

$\frac{2 \cdot 1}{3 \cdot 4} = \frac{2}{12}$ ← Multiply numerators.
← Multiply denominators.

$\frac{2}{12} = \frac{1}{6}$ ← Answer in simplest form

Use the problem $\frac{2}{5} \cdot \frac{3}{4}$ to answer the following questions.

4. When you multiply the numerators, the product is 6.

5. When you multiply the denominators, the product is 20.

6. $\frac{2}{5} \cdot \frac{3}{4} = \frac{6}{20}$

---

## Puzzles, Twisters & Teasers
### 5-7 Itchy Multiplication

Solve each problem and find the answer in the box. Place the letter corresponding to the answer in the blanks to answer the riddle.

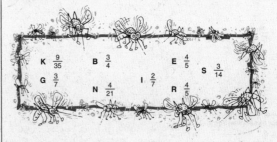

Find the value of each expression if $n = \frac{3}{7}$.

1. The value of $\frac{4}{9}n$    $\frac{4}{21}$   N

2. The value of $\frac{2}{3}n$    $\frac{2}{7}$   I

3. The value of $\frac{3}{5}n$    $\frac{9}{35}$   K

4. The value of $\frac{1}{2}n$    $\frac{3}{14}$   S

What is a mosquito's favorite sport? $\underline{S}\ \underline{K}\ \underline{I}\ \underline{N}$ DIVING
                                     4   3   2   1

## LESSON 5-8 Practice A
### Multiplying Mixed Numbers

Multiply. Write each answer in simplest form.

1. $\frac{1}{2} \cdot 1\frac{1}{3}$
   $\frac{1}{2} \cdot \frac{4}{3}$
   $\frac{2}{3}$

2. $1\frac{1}{5} \cdot \frac{4}{5}$
   $\frac{6}{5} \cdot \frac{4}{5}$
   $\frac{24}{25}$

3. $1\frac{1}{4} \cdot \frac{2}{3}$
   $\frac{5}{4} \cdot \frac{2}{3}$
   $\frac{5}{6}$

4. $1\frac{1}{8} \cdot \frac{2}{5}$
   $\frac{9}{8} \cdot \frac{2}{5}$
   $\frac{9}{20}$

5. $\frac{2}{5} \cdot 1\frac{1}{2}$
   $\frac{2}{5} \cdot \frac{3}{2}$
   $\frac{3}{5}$

6. $1\frac{3}{5} \cdot \frac{1}{3}$
   $\frac{8}{5} \cdot \frac{1}{3}$
   $\frac{8}{15}$

7. $\frac{2}{7} \cdot 1\frac{1}{4}$
   $\frac{5}{14}$

8. $\frac{2}{3} \cdot 1\frac{1}{10}$
   $\frac{11}{15}$

9. $\frac{1}{8} \cdot 1\frac{1}{2}$
   $\frac{3}{16}$

Find each product. Write the answer in simplest form.

10. $\frac{4}{5} \cdot 1\frac{1}{6}$
    $\frac{14}{15}$

11. $\frac{3}{5} \cdot 1\frac{1}{4}$
    $\frac{3}{4}$

12. $1\frac{3}{4} \cdot \frac{1}{3}$
    $\frac{7}{12}$

13. $2 \cdot 1\frac{1}{2}$
    $3$

14. $4 \cdot 2\frac{1}{4}$
    $9$

15. $5 \cdot 1\frac{1}{5}$
    $6$

16. Lin Li makes two and a half dollars per hour baby-sitting her little brother. How much money will she make if she baby-sits for 5 hours?
    $12\frac{1}{2}$ dollars or $12.50

17. Andrea is baking 2 batches of cookies. The recipe calls for $4\frac{1}{2}$ cups of flour for each batch. How many cups of flour will she use?
    9 cups

/17

---

## LESSON 5-8 Practice B
### Multiplying Mixed Numbers

Multiply. Write each answer in simplest form.

1. $1\frac{2}{3} \cdot \frac{4}{5}$
   $1\frac{1}{3}$

2. $1\frac{7}{8} \cdot \frac{4}{5}$
   $1\frac{1}{2}$

3. $2\frac{3}{4} \cdot \frac{1}{5}$
   $\frac{11}{20}$

4. $2\frac{1}{6} \cdot \frac{2}{3}$
   $1\frac{4}{9}$

5. $2\frac{2}{5} \cdot \frac{3}{8}$
   $\frac{9}{10}$

6. $1\frac{3}{4} \cdot \frac{5}{6}$
   $1\frac{11}{24}$

7. $1\frac{1}{6} \cdot \frac{3}{5}$
   $\frac{7}{10}$

8. $\frac{2}{9} \cdot 2\frac{1}{7}$
   $\frac{10}{21}$

9. $2\frac{3}{11} \cdot \frac{7}{10}$
   $1\frac{13}{22}$

Find each product. Write the answer in simplest form.

10. $\frac{6}{7} \cdot 1\frac{1}{4}$
    $1\frac{1}{14}$

11. $\frac{5}{8} \cdot 1\frac{3}{5}$
    $1$

12. $2\frac{4}{9} \cdot \frac{1}{6}$
    $\frac{11}{27}$

13. $1\frac{3}{10} \cdot 1\frac{1}{3}$
    $1\frac{11}{15}$

14. $2\frac{1}{2} \cdot 2\frac{1}{2}$
    $6\frac{1}{4}$

15. $1\frac{2}{3} \cdot 3\frac{1}{2}$
    $5\frac{5}{6}$

16. Dominick lives $1\frac{3}{4}$ miles from his school. If his mother drives him half the way, how far will Dominick have to walk to get to school?
    $\frac{7}{8}$ mile

17. Katoni bought $2\frac{1}{2}$ dozen donuts to bring to the office. Since there are 12 donuts in a dozen, how many donuts did Katoni buy?
    30 donuts

---

## LESSON 5-8 Practice C
### Multiplying Mixed Numbers

Multiply. Write each answer in simplest form. Do NOT accept improper fractions

1. $\frac{5}{9} \cdot 2\frac{2}{7}$
   $1\frac{17}{63}$

2. $1\frac{11}{12} \cdot \frac{6}{7}$
   $1\frac{9}{14}$ 23/14

3. $2\frac{4}{9} \cdot \frac{7}{8}$
   $2\frac{5}{36}$ 77/36

4. $3\frac{2}{3} \cdot \frac{3}{5}$
   $2\frac{1}{5}$ 11/5

5. $1\frac{13}{14} \cdot 1\frac{1}{4}$
   $1\frac{5}{8}$ 13/8

6. $2\frac{3}{10} \cdot \frac{5}{6}$
   $1\frac{11}{12}$ 23/12

7. $1\frac{7}{8} \cdot \frac{3}{5}$
   $1\frac{1}{8}$

8. $3\frac{2}{7} \cdot \frac{3}{10}$
   $\frac{69}{70}$

9. $4\frac{2}{3} \cdot \frac{8}{9}$
   $4\frac{4}{27}$ 112/27

Find each product. Write the answer in simplest form.

10. $\frac{10}{11} \cdot 3\frac{3}{7} \cdot 2$
    $6\frac{18}{77}$ 480/77

11. $2\frac{4}{7} \cdot \frac{4}{5} \cdot 1\frac{1}{2}$
    $3\frac{3}{35}$ 108/35

12. $\frac{9}{12} \cdot 2\frac{5}{8} \cdot 3\frac{1}{4}$
    $6\frac{27}{80}$ 507/80

13. $6\frac{1}{5} \cdot 10 \cdot 3\frac{5}{8}$
    $235\frac{3}{5}$ 1178/5

14. $1\frac{7}{9} \cdot \frac{2}{5} \cdot 5\frac{1}{10}$
    $3\frac{47}{75}$ 272/75

15. $2\frac{6}{7} \cdot 1\frac{8}{9} \cdot \frac{7}{8}$
    $4\frac{13}{18}$ 85/18

Evaluate each expression.

16. $\frac{3}{4} \cdot c$ for $c = 4\frac{4}{5}$
    $3\frac{3}{5}$ 18/5

17. $1\frac{3}{10} \cdot x$ for $x = 2\frac{2}{3}$
    $3\frac{7}{15}$ 52/15

18. $\frac{2}{9} \cdot h$ for $h = 3\frac{5}{6}$
    $\frac{23}{27}$

19. $\frac{3}{4} \cdot q$ for $q = 2\frac{7}{8}$
    $2\frac{5}{32}$ 69/32

20. A train travels at $110\frac{3}{10}$ miles per hour. At this rate, how far will the train travel in $2\frac{1}{2}$ hours?
    $275\frac{3}{4}$ miles  1103/4

21. A sandbox is $1\frac{1}{2}$ feet tall, $1\frac{5}{8}$ feet wide, and $4\frac{1}{2}$ feet long. How many cubic feet of sand is needed to fill the box? (Volume = length • width • height)
    $9\frac{3}{4}$ cubic feet of sand

39/4

---

## LESSON 5-8 Reteach
### Multiplying Mixed Numbers

To find $\frac{1}{3}$ of $2\frac{1}{2}$, first change $2\frac{1}{2}$ to an improper fraction.

$2\frac{1}{2} = \frac{5}{2}$

Then multiply as you would with two proper fractions.

Check to see if you can divide by the GCF to make the problem simpler. Then multiply the numerators and multiply the denominators.

The problem is now $\frac{1}{3} \cdot \frac{5}{2}$.

$\frac{1 \cdot 5}{3 \cdot 2} = \frac{5}{6}$

So, $\frac{1}{3} \cdot 2\frac{1}{2}$ is $\frac{5}{6}$.

/30

Rewrite each mixed number as an improper fraction. Is it possible to simplify before you multiply? If so, what is the GCF?

1. $\frac{1}{4} \cdot 1\frac{1}{3}$
   $= \frac{1}{4} \cdot \frac{4}{3}$
   $\frac{1}{3}$

2. $\frac{1}{6} \cdot 2\frac{1}{2}$
   $= \frac{1}{6} \cdot \frac{5}{2}$
   $\frac{5}{12}$

3. $\frac{1}{8} \cdot 1\frac{1}{2}$
   $= \frac{1}{8} \cdot \frac{3}{2}$
   $\frac{3}{16}$

4. $\frac{1}{3} \cdot 1\frac{2}{5}$
   $= \frac{1}{3} \cdot \frac{7}{5}$
   $\frac{7}{15}$

5. $1\frac{1}{3} \cdot 1\frac{2}{3}$
   $\frac{4}{3} \cdot \frac{5}{3}$
   $2\frac{2}{9}$

6. $1\frac{1}{2} \cdot 1\frac{1}{3}$
   $\frac{3}{2} \cdot \frac{4}{3}$
   $2$

7. $1\frac{3}{4} \cdot 2\frac{1}{2}$
   $\frac{7}{4} \cdot \frac{5}{2}$
   $4\frac{3}{8}$

8. $1\frac{1}{6} \cdot 2\frac{2}{3}$
   $\frac{7}{6} \cdot \frac{8}{3}$
   $3\frac{1}{9}$

9. $3\frac{1}{3} \cdot \frac{2}{5}$
   $1\frac{1}{3}$

10. $2\frac{1}{2} \cdot \frac{1}{5}$
    $\frac{1}{2}$

11. $1\frac{3}{4} \cdot 2\frac{1}{2}$
    $4\frac{3}{8}$

12. $3\frac{1}{3} \cdot 1\frac{1}{5}$
    $4$

/24

## Challenge
### 5-8 And They're Off!

Like many sports, horse racing uses a special system of measurement. Horse races are measured in units called *furlongs*. One furlong equals $\frac{1}{8}$ mile. The races described below have different furlong lengths, but they all offer the same prize money to their winners—$1,000,000!

**Write the length in miles of each of these horse races in simplest form.**

1. Santa Anita Derby, California

   Race Length: 9 furlongs    Length in Miles: $1\frac{1}{8}$ miles

2. Kentucky Derby, Kentucky

   Race Length: 10 furlongs   Length in Miles: $1\frac{1}{4}$ miles

3. Preakness Stakes, Maryland

   Race Length: $9\frac{1}{2}$ furlongs   Length in Miles: $1\frac{3}{16}$ miles

4. Belmont Stakes, New York

   Race Length: 12 furlongs   Length in Miles: $1\frac{1}{2}$ miles

5. Breeders' Cup Juvenile, New York

   Race Length: $8\frac{1}{2}$ furlongs   Length in Miles: $1\frac{1}{16}$ miles

---

## Problem Solving
### 5-8 Multiplying Mixed Numbers

**Use the recipe to answer the questions.**

| CHOCOLATE CHIP COOKIES |
|---|
| Servings: 1 batch |
| $1\frac{2}{3}$ cups flour |
| $\frac{3}{4}$ teaspoon baking soda |
| $\frac{1}{2}$ cup white sugar |
| $2\frac{1}{3}$ cups semisweet chocolate chips |
| $\frac{1}{2}$ cup brown sugar |
| $\frac{3}{4}$ cup butter |
| 1 egg |
| $1\frac{1}{4}$ teaspoons vanilla |

1. If you want to make $2\frac{1}{2}$ batches, how much flour would you need?
   $4\frac{1}{6}$ cups

2. If you want to make only $1\frac{1}{2}$ batches, how much chocolate chips would you need?
   $3\frac{1}{2}$ cups

3. You want to bake $3\frac{1}{4}$ batches. How much vanilla do you need in all?
   $4\frac{1}{16}$ teaspoons

**Choose the letter for the best answer.**

4. If you make $1\frac{1}{4}$ batches, how much baking soda would you need?

   A $\frac{3}{16}$ teaspoon   C $\frac{3}{5}$ teaspoon
   B $\frac{5}{16}$ teaspoon   D $\frac{15}{16}$ teaspoon

5. How many cups of white sugar do you need to make $3\frac{1}{2}$ batches of cookies?

   F $3\frac{1}{2}$ cups   H $1\frac{1}{2}$ cups
   (G) $1\frac{3}{4}$ cups   J $1\frac{1}{4}$ cups

6. Dan used $2\frac{1}{4}$ cups of butter to make chocolate chip cookies using the above recipe. How many batches of cookies did he make?

   (A) 3 batches   C 5 batches
   B 4 batches    D 6 batches

7. One bag of chocolate chips holds 2 cups. If you buy five bags, how many cups of chips will you have left over after baking $2\frac{1}{2}$ batches of cookies?

   (F) $4\frac{1}{6}$ cups   H $2\frac{1}{3}$ cups
   G $5\frac{5}{6}$ cups   J $\frac{1}{3}$ cup

---

## Reading Strategies
### 5-8 Use a Flow Chart

Mixed Number

Whole number → $2\frac{1}{2}$ ← Fraction

Improper Fraction

$\frac{5}{2}$

You can change mixed numbers to improper fractions.

$2\frac{1}{2}$ = 5 halves or $\frac{5}{2}$ ← improper fraction

1. What is the mixed number in the above example? $2\frac{1}{2}$
2. What is the improper fraction? $\frac{5}{2}$
3. How many halves are in $2\frac{1}{2}$? 5

**Use the flowchart below to help you change a mixed number to an improper fraction.**

| Multiply the denominator by the whole number. | → | Add the numerator. | → | The denominator stays the same. |

**Change $3\frac{2}{5}$ to an improper fraction.**

4. What is the first step? Multiply 5 • 3.
5. What is the next step? Add 2 to 15.
6. The improper fraction is $\frac{17}{5}$

---

## Puzzles, Twisters & Teasers
### 5-8 All Mixed Up!

Rami was carrying a set of cards, but he tripped. The cards fell on the floor and are all mixed up. Help Rami put them in order by solving each problem.

Once you have solved the problems, place the cards in order from least to greatest. When in order, the letters will spell out a message!

The message is... GOOD JOB

## Practice A
### 5-9 Dividing Fractions and Mixed Numbers

**Find the reciprocal.**

1. $\frac{1}{2}$ — 2
2. $\frac{2}{3}$ — $\frac{3}{2}$
3. $\frac{1}{5}$ — 5
4. $\frac{1}{3}$ — 3
5. $\frac{3}{5}$ — $\frac{5}{3}$
6. $1\frac{1}{4}$ — $\frac{4}{5}$
7. $\frac{2}{5}$ — $\frac{5}{2}$
8. $\frac{3}{7}$ — $\frac{7}{3}$
9. $1\frac{1}{2}$ — $\frac{2}{3}$

**Divide. Write each answer in simplest form.**

10. $\frac{2}{3} \div 2$ = $\frac{2}{3} \cdot \frac{1}{2}$ = $\frac{1}{3}$
11. $\frac{1}{2} \div \frac{3}{4}$ = $\frac{1}{2} \cdot \frac{4}{3}$ = $\frac{2}{3}$
12. $\frac{5}{6} \div \frac{1}{4}$ = $\frac{5}{6} \cdot \frac{4}{1}$ = $3\frac{1}{3}$
13. $\frac{3}{5} \div \frac{1}{5}$ = $\frac{3}{5} \cdot \frac{5}{1}$ = 3
14. $\frac{7}{9} \div 3$ = $\frac{7}{9} \cdot \frac{1}{3}$ = $\frac{7}{27}$
15. $1\frac{1}{2} \div \frac{1}{2}$ = $1\frac{1}{2} \cdot \frac{2}{1}$ = 3

16. Stella has 6 pounds of chocolate. She will use $\frac{2}{3}$ pound of the chocolate to make one cake. How many cakes can she make? — **9 cakes**
17. Todd has $\frac{8}{9}$ pound of clay. He will use $\frac{1}{3}$ pound to make each action figure. How many action figures can he make? — **2 action figures**
18. Dylan gives his two guinea pigs a total of $\frac{3}{4}$ cup of food every day. If each guinea pig gets the same amount of food, how much do they each get each day? — **$\frac{3}{8}$ cup of food**

## Practice B
### 5-9 Dividing Fractions and Mixed Numbers

**Find the reciprocal.**

1. $\frac{5}{7}$ — $\frac{7}{5}$
2. $\frac{9}{8}$ — $\frac{8}{9}$
3. $\frac{3}{5}$ — $\frac{5}{3}$
4. $\frac{1}{10}$ — 10
5. $\frac{4}{9}$ — $\frac{9}{4}$
6. $\frac{13}{14}$ — $\frac{14}{13}$
7. $1\frac{1}{3}$ — $\frac{3}{4}$
8. $2\frac{4}{5}$ — $\frac{5}{14}$
9. $3\frac{1}{6}$ — $\frac{6}{19}$

**Divide. Write each answer in simplest form.**

10. $\frac{5}{6} \div 5$ = $\frac{1}{6}$
11. $2\frac{3}{4} \div 1\frac{4}{7}$ = $1\frac{3}{4}$
12. $\frac{7}{8} \div \frac{2}{3}$ = $1\frac{5}{16}$
13. $3\frac{1}{4} \div 2\frac{3}{4}$ = $1\frac{2}{11}$
14. $\frac{9}{10} \div 3$ = $\frac{3}{10}$
15. $\frac{3}{4} \div 9$ = $\frac{1}{12}$
16. $2\frac{6}{9} \div \frac{6}{7}$ = $3\frac{1}{9}$
17. $\frac{5}{6} \div 2\frac{3}{10}$ = $\frac{25}{69}$
18. $2\frac{1}{8} \div \frac{3}{4}$ = $\frac{17}{26}$

19. The rope in the school gymnasium is $10\frac{1}{2}$ feet long. To make it easier to climb, the gym teacher tied a knot in the rope every $\frac{3}{4}$ foot. How many knots are in the rope? — **14 knots**
20. Mr. Fulton bought $12\frac{1}{2}$ pounds of ground beef for the cookout. He plans on using $\frac{1}{4}$ pound of beef for each hamburger. How many hamburgers can he make? — **50 hamburgers**
21. Mrs. Marks has $9\frac{1}{4}$ ounces of fertilizer for her plants. She plans on using $\frac{3}{4}$ ounce of fertilizer for each plant. How many plants can she fertilize? — **12 plants**

## Practice C
### 5-9 Dividing Fractions and Mixed Numbers

**Find the reciprocal.**

1. $10\frac{1}{2}$ — $\frac{2}{21}$
2. $6\frac{3}{7}$ — $\frac{7}{45}$
3. $2\frac{8}{9}$ — $\frac{9}{26}$
4. $15\frac{1}{4}$ — $\frac{4}{61}$
5. $9\frac{2}{3}$ — $\frac{3}{29}$
6. $7\frac{5}{8}$ — $\frac{8}{61}$

**Divide. Write each answer in simplest form.**

7. $\frac{8}{10} \div 1\frac{5}{6}$ = $\frac{24}{55}$
8. $\frac{8}{9} \div \frac{6}{7}$ = $1\frac{1}{27}$
9. $3\frac{3}{5} \div 2\frac{1}{4}$ = $1\frac{3}{5}$
10. $4\frac{1}{2} \div 2\frac{3}{8}$ = $1\frac{17}{19}$
11. $5\frac{5}{6} \div 3\frac{1}{6}$ = $1\frac{16}{19}$
12. $\frac{11}{12} \div 2\frac{5}{8}$ = $\frac{22}{63}$
13. $1\frac{9}{13} \div \frac{3}{8}$ = $4\frac{20}{39}$
14. $6\frac{4}{5} \div 3\frac{2}{9}$ = $2\frac{16}{145}$
15. $8\frac{2}{11} \div 2\frac{4}{7}$ = $3\frac{2}{11}$
16. $9\frac{6}{13} \div 10$ = $\frac{123}{130}$
17. $12\frac{1}{3} \div 5\frac{4}{5}$ = $2\frac{11}{87}$
18. $9\frac{2}{3} \div 6\frac{8}{9}$ = $1\frac{25}{62}$

19. The area of the public swimming pool is $510\frac{7}{8}$ square feet. The pool is $30\frac{1}{2}$ feet long. What is the width of the pool? — **$16\frac{3}{4}$ feet**
20. At the bank, Pamela exchanged all of her quarters for 16 five-dollar bills. How many quarters did Pamela exchange? — **320 quarters**
21. Barbara has $16\frac{1}{5}$ yards of fabric. She will use $5\frac{2}{5}$ yards to make each costume. How many costumes can Barbara make? — **3 costumes**

## Reteach
### 5-9 Dividing Fractions and Mixed Numbers

Two numbers are reciprocals if their product is 1. $\frac{2}{3}$ and $\frac{3}{2}$ are reciprocals because $\frac{2}{3} \cdot \frac{3}{2} = \frac{6}{6} = 1$.

Dividing by a fraction is the same as multiplying by its reciprocal.
$\frac{1}{4} \div 2 = \frac{1}{8}$  $\frac{1}{4} \cdot \frac{1}{2} = \frac{1}{8}$

So, you can use reciprocals to divide by fractions.

To find $\frac{2}{3} \div 4$, first rewrite the expression as a multiplication expression using the reciprocal of the divisor, 4.
$\frac{2}{3} \cdot \frac{1}{4}$

Then use canceling to find the product in simplest form.
$\frac{2}{3} \div 4 = \frac{2}{3} \cdot \frac{1}{4} = \frac{1}{3} \cdot \frac{1}{2} = \frac{1}{6}$

To find $3\frac{1}{4} \div 1\frac{1}{2}$, first rewrite the expression using improper fractions.
$\frac{13}{4} \div \frac{3}{2}$

Next, write the expression as a multiplication expression.
$\frac{13}{4} \cdot \frac{2}{3}$

$3\frac{1}{4} \div 1\frac{1}{2} = \frac{13}{4} \div \frac{3}{2} = \frac{13}{4} \cdot \frac{2}{3} = \frac{13}{2} \cdot \frac{1}{3} = \frac{13}{6} = 2\frac{1}{6}$

**Divide. Write each answer in simplest form.**

1. $\frac{1}{4} \div 3$ = $\frac{1}{4} \div \frac{3}{1}$ = $\frac{1}{4} \cdot \frac{1}{3}$ = $\frac{1}{12}$
2. $1\frac{1}{2} \div 1\frac{1}{4}$ = $\frac{3}{2} \div \frac{5}{4}$ = $\frac{3}{2} \cdot \frac{4}{5}$ = $1\frac{1}{5}$
3. $\frac{3}{8} \div \frac{2}{1}$ = $\frac{3}{8} \cdot \frac{1}{2}$ = $\frac{3}{16}$
4. $2\frac{1}{3} \div 1\frac{3}{4}$ = $\frac{7}{3} \div \frac{7}{4}$ = $\frac{7}{3} \cdot \frac{4}{7}$ = $1\frac{1}{3}$

5. $\frac{1}{5} \div 2$ = $\frac{1}{10}$
6. $1\frac{1}{6} \div 2\frac{2}{3}$ = $\frac{7}{16}$
7. $\frac{1}{8} \div 4$ = $\frac{1}{32}$
8. $3\frac{1}{2} \div \frac{1}{2}$ = $6\frac{1}{4}$

Holt Mathematics

## LESSON 5-9 Challenge
### Inching Across the U.S.A.

You can use a map and a ruler to find the distance between places. On the map below, for example, you measure that 2 inches separate Kansas City, Missouri, and Richmond, Virginia. The map scale shows that $\frac{1}{4}$ inch on the map equals 140 miles. So the distance between Kansas City and Richmond is 1,120 miles.

Calculations:
$2 \div \frac{1}{4} = 8$
$8 \times 140 = 1,120$

**Use the map and a ruler to find the distance in miles between each pair of cities.**

1. Miami, Florida, and New Orleans, Louisiana  __770 miles__
2. Denver, Colorado, and Los Angeles, California  __1,015 miles__
3. Seattle, Washington, and Minneapolis, Minnesota  __1,540 miles__
4. Washington, D.C., and Atlanta, Georgia  __665 miles__
5. Oklahoma City, Oklahoma, and Pittsburgh, Pennsylvania  __1,120 miles__
6. San Francisco, California, and Boston, Massachusetts  __3,080 miles__

$\frac{1}{4}$ in. = 140 mi.

---

## LESSON 5-9 Problem Solving
### Dividing Fractions and Mixed Numbers

**Write the correct answer in simplest form.**

1. Horses are measured in units called *hands*. One inch equals $\frac{1}{4}$ hand. The average Clydesdale horse is $17\frac{1}{5}$ hands high. What is the horse's height in inches? in feet?
   $68\frac{4}{5}$ inches; $5\frac{11}{15}$ feet

2. Cloth manufacturers use a unit of measurement called a *finger*. One finger is equal to $4\frac{1}{2}$ inches. If 25 inches are cut off a bolt of cloth, how many fingers of cloth were cut?
   $5\frac{5}{9}$ fingers

3. People in England measure weights in units called *stones*. One pound equals $\frac{1}{14}$ of a stone. If a cat weighs $\frac{3}{4}$ stone, how many pounds does it weigh?
   $10\frac{1}{2}$ pounds

4. The hiking trail is $\frac{9}{10}$ mile long. There are 6 markers evenly posted along the trail to direct hikers. How far apart are the markers placed?
   $\frac{3}{20}$ mile

**Choose the letter for the best answer.**

5. A cake recipe calls for $1\frac{1}{2}$ cups of butter. One tablespoon equals $\frac{1}{16}$ cup. How many tablespoons of butter do you need to make the cake?
   **A** 24 tablespoons
   **B** 8 tablespoons
   **C** $\frac{3}{32}$ tablespoon
   **D** 9 tablespoons

6. Printed letters are measured in units called *points*. One point equals $\frac{1}{72}$ inch. If you want the title of a paper you are typing on a computer to be $\frac{1}{2}$ inch tall, what type point size should you use?
   **F** 144 point
   **G** 36 point
   **H** $\frac{1}{36}$ point
   **J** $\frac{1}{144}$ point

7. Phyllis bought 14 yards of material to make chair cushions. She cut the material into pieces $1\frac{3}{4}$ yards long to make each cushion. How many cushions did Phyllis make?
   **A** 4 cushions
   **C** 8 cushions
   **B** 6 cushions
   **D** $24\frac{1}{2}$ cushions

8. Dry goods are sold in units called *pecks* and *bushels*. One peck equals $\frac{1}{4}$ bushel. If Peter picks $5\frac{1}{2}$ bushels of peppers, how many pecks of peppers did Peter pick?
   **F** $1\frac{3}{8}$ pecks
   **H** 20 pecks
   **G** 11 pecks
   **J** 22 pecks

---

## LESSON 5-9 Reading Strategies
### Using Models

Fraction bars help you picture dividing by fractions.

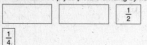

In the problem $2\frac{1}{2} \div \frac{1}{4}$, think: How many one-fourths are there in $2\frac{1}{2}$?

**Use the picture to answer each question.**

1. Count the number of $\frac{1}{4}$'s in the fraction bars above. How many are there?  __10__
2. $2\frac{1}{2} \div \frac{1}{4} = $  __10__

In the problem $2\frac{1}{2} \times 4$, think $2\frac{1}{2}$ four times.

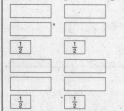

**Use the picture to answer each question.**

3. How many whole fraction bars are there?  __8__
4. How many one-half fraction bars are there?  __4__
5. When you add the whole bars and half bars together you get __10__ whole bars.
6. Compare the multiplication and division examples. What do you notice about the answer you get when you divide by $\frac{1}{4}$ or multiply by 4?
   __The answer is the same.__

---

## LESSON 5-9 Puzzles, Twisters & Teasers
### Divide and Conquer!

You've heard there is a Pot of Gold to be found. Begin at "S"(start). Decide whether the first statement is true or false. Circle your answer, and move as directed. Go to problem 2. Decide whether the statement is true or false and move as directed. Can you conquer the maze and make it to the Pot of Gold?

1. The reciprocal of $\frac{5}{7}$ is $\frac{7}{5}$.
   (True:) 7 steps right and 6 steps down.
   False: 6 steps right and 7 steps down.

2. $\frac{2}{3} \div \frac{1}{3} = \frac{2}{9}$
   True: 3 steps right and 7 steps up.
   (False:) 7 steps right and 3 steps up.

3. The reciprocal of $4\frac{3}{5}$ is $4\frac{5}{3}$.
   True: 6 steps right and 3 steps down.
   (False:) 3 steps right and 6 steps down.

4. $3\frac{1}{2} + 2\frac{1}{6} = 5\frac{2}{3}$
   (True:) 2 steps diagonally up and to the right.
   False: 2 steps diagonally down and to the right.

## Practice A
### 5-10 Solving Fraction Equations: Multiplication and Division

Solve each equation. Write the answer in simplest form.

1. $\frac{1}{2}x = 2$
   $x = 4$

2. $2t = \frac{2}{3}$
   $t = \frac{1}{3}$

3. $\frac{1}{3}a = 3$
   $a = 9$

4. $\frac{r}{2} = 4$
   $r = 8$

5. $\frac{b}{3} = 6$
   $b = 18$

6. $2y = \frac{1}{5}$
   $y = \frac{1}{10}$

7. $\frac{1}{4}d = 2$
   $d = 8$

8. $\frac{b}{5} = 6$
   $b = 30$

9. $\frac{q}{10} = \frac{1}{5}$
   $q = 2$

10. $\frac{1}{3}s = 4$
    $s = 12$

11. $\frac{h}{2} = 2$
    $h = 4$

12. $\frac{1}{4}c = 1$
    $c = 4$

Circle the correct answer.

13. Tate earned $9 for working $\frac{3}{4}$ of an hour. Which equation can be used to find Tate's hourly rate?

    A $9h = \frac{3}{4}$
    B $9 + \frac{3}{4} = h$
    C $\frac{3}{4}h = 9$
    D $9 - \frac{3}{4} = h$

14. Which operation should you use to solve the equation $5x = 2$?
    F addition
    G subtraction
    H multiplication
    J division

15. A number $n$ is divided by 2, and the quotient is $\frac{1}{3}$. Write an equation to model this problem.
    $\frac{n}{2} = \frac{1}{3}$

16. A number $n$ is multiplied by $\frac{1}{4}$, and the product is 5. Write and solve an equation to model this problem.
    $\frac{1}{4}n = 5; n = 20$

## Practice B
### 5-10 Solving Fraction Equations: Multiplication and Division

Solve each equation. Write the answer in simplest form. Check your answers.

1. $\frac{1}{4}x = 6$
   $x = 24;$
   $\frac{1}{4} \cdot 24 = 6$ ✓

2. $2t = \frac{4}{7}$
   $t = \frac{2}{7};$
   $2 \cdot \frac{2}{7} = \frac{4}{7}$ ✓

3. $\frac{3}{5}a = 3$
   $a = 5;$
   $\frac{3}{5} \cdot 5 = 3$ ✓

4. $\frac{r}{6} = 8$
   $r = 48;$
   $\frac{48}{6} = 8$ ✓

5. $\frac{2b}{9} = 4$
   $b = 18;$
   $\frac{(2 \cdot 18)}{9} = 4$ ✓

6. $3y = \frac{4}{5}$
   $y = \frac{4}{15};$
   $3 \cdot \frac{4}{15} = \frac{4}{5}$ ✓

7. $\frac{2}{3}d = 5$
   $d = 7\frac{1}{2};$
   $\frac{2}{3} \cdot 7\frac{1}{2} = 5$ ✓

8. $2f = \frac{1}{6}$
   $f = \frac{1}{12};$
   $2 \cdot \frac{1}{12} = \frac{1}{6}$ ✓

9. $4q = \frac{2}{9}$
   $q = \frac{1}{18};$
   $4 \cdot \frac{1}{18} = \frac{2}{9}$ ✓

10. $\frac{1}{2}s = 2$
    $s = 4;$
    $\frac{1}{2} \cdot 4 = 2$ ✓

11. $\frac{h}{7} = 5$
    $h = 35;$
    $\frac{35}{7} = 5$ ✓

12. $\frac{1}{4}c = 9$
    $c = 36;$
    $\frac{1}{4} \cdot 36 = 9$ ✓

13. $5g = \frac{5}{6}$
    $g = \frac{1}{6};$
    $5 \cdot \frac{1}{6} = \frac{5}{6}$ ✓

14. $3k = \frac{1}{9}$
    $k = \frac{1}{27};$
    $3 \cdot \frac{1}{27} = \frac{1}{9}$ ✓

15. $\frac{3x}{5} = 6$
    $x = 10;$
    $\frac{(3 \cdot 10)}{5} = 6$ ✓

16. It takes 3 buckets of water to fill $\frac{1}{3}$ of a fish tank. How many buckets are needed to fill the whole tank?
    9 buckets

17. Jenna got 12, or $\frac{3}{5}$, of her answers on the test right. How many questions were on the test?
    20 questions

18. It takes Charles 2 minutes to run $\frac{1}{4}$ of a mile. How long will it take Charles to run a mile?
    8 minutes

## Practice C
### 5-10 Solving Fraction Equations: Multiplication and Division

Solve each equation. Write the answer in simplest form. Check your answers.

1. $\frac{2}{3}x = 10$
   $x = 15;$
   $\frac{2}{3} \cdot 15 = 10$ ✓

2. $5t = \frac{10}{15}$
   $t = \frac{2}{15};$
   $5 \cdot \frac{2}{15} = \frac{10}{15}$ ✓

3. $\frac{6}{7}a = 9$
   $a = 10\frac{1}{2};$
   $\frac{6}{7} \cdot 10\frac{1}{2} = 9$ ✓

4. $\frac{r}{11} = 12$
   $r = 132;$
   $\frac{132}{11} = 12$ ✓

5. $\frac{6b}{9} = 15$
   $b = 22\frac{1}{2};$
   $\frac{(6 \cdot 22\frac{1}{2})}{9} = 15$ ✓

6. $7y = \frac{7}{8}$
   $y = \frac{1}{8};$
   $7 \cdot \frac{1}{8} = \frac{7}{8}$ ✓

7. $\frac{4}{5}d = 15$
   $d = 18\frac{3}{4};$
   $\frac{4}{5} \cdot 18\frac{3}{4} = 15$ ✓

8. $4f = \frac{1}{9}$
   $f = \frac{1}{36};$
   $4 \cdot \frac{1}{36} = \frac{1}{9}$ ✓

9. $7q = \frac{3}{5}$
   $q = \frac{3}{35};$
   $7 \cdot \frac{3}{35} = \frac{3}{5}$ ✓

10. $\frac{7}{8}s = 14$
    $s = 16;$
    $\frac{7}{8} \cdot 16 = 14$ ✓

11. $\frac{h}{12} = 6$
    $h = 72;$
    $\frac{72}{12} = 6$ ✓

12. $\frac{3}{10}c = \frac{2}{3}$
    $c = 2\frac{2}{9};$
    $\frac{3}{10} \cdot 2\frac{2}{9} = \frac{2}{3}$ ✓

13. $\frac{5g}{6} = \frac{7}{12}$
    $g = \frac{7}{10};$
    $\frac{(5 \cdot \frac{7}{10})}{6} = \frac{7}{12}$ ✓

14. $\frac{3k}{9} = \frac{5}{6}$
    $k = 2\frac{1}{2};$
    $\frac{(3 \cdot 2\frac{1}{2})}{9} = \frac{5}{6}$ ✓

15. $5\frac{1}{2}n = 3$
    $n = \frac{6}{11};$
    $5\frac{1}{2} \cdot \frac{6}{11} = 3$ ✓

16. Anya worked $8\frac{1}{4}$ hours on Saturday and $6\frac{1}{4}$ hours on Sunday. She earned a total of $137.75 for both days combined. How much does Anya make per hour?
    $9.50

17. Ernest rode his bike $6\frac{1}{4}$ miles on Saturday and $8\frac{1}{2}$ miles on Sunday. He rode for a total of $88\frac{1}{2}$ minutes for both days combined. How long does it take him to ride a mile on his bike?
    6 minutes

## Reteach
### 5-10 Solving Fraction Equations: Multiplication and Division

You can write related facts using multiplication and division.
$3 \cdot 4 = 12 \qquad 4 = 12 \div 3$

You can use related facts to solve equations.

A. $\frac{2}{3} \cdot x = 12$
   Think: $12 \div \frac{2}{3} = x$
   $x = 12 \cdot \frac{3}{2}$
   $x = \frac{12}{1} \cdot \frac{3}{2}$
   $x = \frac{36}{2}$
   $x = 18$

   Check: $\frac{2}{3} \cdot x = 12$
   $\frac{2}{3} \cdot 18 \stackrel{?}{=} 12$ Substitute
   $\frac{2}{3} \cdot \frac{18}{1} \stackrel{?}{=} 12$
   $\frac{36}{3} \stackrel{?}{=} 12$
   $12 = 12$ ✓

B. $\frac{2x}{5} = 3$
   $\frac{2}{5} \cdot x = 3$
   Think: $3 \div \frac{2}{5} = x$
   $x = 3 \cdot \frac{5}{2}$
   $x = \frac{3}{1} \cdot \frac{5}{2}$
   $x = \frac{15}{2}$
   $x = 7\frac{1}{2}$

   Check: $\frac{2x}{5} = 3$
   $\frac{2}{5} \cdot x \stackrel{?}{=} 3$
   $\frac{2}{5} \cdot \frac{15}{2} \stackrel{?}{=} 3$ Substitute
   $\frac{30}{10} \stackrel{?}{=} 3$
   $3 = 3$ ✓

Use related facts to solve each equation. Then check each answer.

1. $\frac{1}{4} \cdot x = 3$
   $x = 12$ ✓

2. $\frac{3x}{4} = 2$
   $x = 2\frac{2}{3}$ ✓

3. $\frac{3}{5} \cdot x = \frac{2}{3}$
   $x = 1\frac{1}{9}$ ✓

4. $\frac{1}{3} \cdot x = 6$
   $x = 18$ ✓

5. $\frac{2x}{5} = 1$
   $x = 2\frac{1}{2}$ ✓

6. $\frac{1}{8} \cdot x = 3$
   $x = 24$ ✓

## Challenge
### 5-10 Crawly Creature Equations

A millipede called the *Illacme plenipes* holds the record for the creature with the most legs—750! However, most millipedes have only 30 legs. Shown below are some other many-legged creatures.

Let  = the number of legs most millipedes have. Use this information to solve the equations and find how many legs each other crawly creature has.

$\frac{8}{15} \cdot$ 🐛 = 🐛

$\frac{3}{5} \cdot$ 🐛 $\cdot \frac{5}{6}$ = 🕷

$\frac{3}{4} \cdot$ 🕷 = 🪰

🐛 $\cdot \frac{1}{3}$ = 🦀

Caterpillars    Spiders    Insects    Crabs

16 legs    8 legs    6 legs    10 legs

## Problem Solving
### 5-10 Solving Fraction Equations: Multiplication and Division

Solve.

1. The number of T-shirts is multiplied by $\frac{1}{2}$ and the product is 18. Write and solve an equation for the number of T-shirts, where *t* represents the number of T-shirts.

   $t \cdot \frac{1}{2} = 18; t = 36$

2. The number of students is divided by 18 and the quotient is $\frac{1}{6}$. Write and solve an equation for the number of students, where *s* represents the number of students.

   $s \div 18 = \frac{1}{6}; s = 3$

3. The number of players is multiplied by $2\frac{1}{2}$ and the product is 25. Write and solve an equation for the number of players, where *p* represents the number of players.

   $p \cdot 2\frac{1}{2} = 25; p = 10$

4. The number of chairs is divided by $\frac{1}{4}$ and the quotient is 12. Write and solve an equation for the number of chairs, where *c* represents the number of chairs.

   $c \div \frac{1}{4} = 12; c = 48$

Circle the letter of the correct answer.

5. Paco bought 10 feet of rope. He cut it into several $\frac{5}{6}$-foot pieces. Which equation can you use to find how many pieces of rope Paco cut?

   A $\frac{5}{6} \div 10 = x$
   B $\frac{5}{6} \div x = 10$
   **C** $10 \div x = \frac{5}{6}$
   D $10x = \frac{5}{6}$

6. Each square on the graph paper has an area of $\frac{4}{9}$ square inch. What is the length and width of each square?

   F $\frac{1}{9}$ inch
   **G** $\frac{2}{3}$ inch
   H $\frac{2}{9}$ inch
   J $\frac{1}{3}$ inch

7. Which operation should you use to solve the equation $6x = \frac{3}{8}$?

   A addition
   B subtraction
   C multiplication
   **D** division

8. A fraction divided by $\frac{2}{3}$ is equal to $1\frac{1}{4}$. What is that fraction?

   F $\frac{1}{3}$
   **G** $\frac{5}{6}$
   H $\frac{1}{4}$
   J $\frac{1}{2}$

## Reading Strategies
### 5-10 Compare and Contrast

When you compare two or more things, you look at how they are alike and how they are different. Equations with fractions follow the same rules as equations with whole numbers. Compare the whole number equation with the fraction equation.

| | Whole Number Equation | Fraction Equation |
|---|---|---|
| **Step 1:** Divide both sides of the equation by the same number to get the variable by itself. | $3y = 36$ | $\frac{3}{4}y = 6$ |
| **Step 2:** Divide on both sides of the equation. | $\frac{3}{3}y = \frac{36}{3}$ | $\frac{4}{3} \cdot \frac{3}{4}y = 6 \cdot \frac{4}{3}$ ← Dividing by $\frac{3}{4}$ is the same as multiplying by $\frac{4}{3}$. Multiply 6 by $\frac{4}{3}$. |
| | | $y = 6 \cdot \frac{4}{3}$ |
| | | $y = \frac{24}{3}$ |
| **Step 3:** Simplify. | $y = 12$ | $y = 8$ |

Compare the equations and answer each question.

1. What is the same about solving equations with whole numbers and solving equations with fractions?

   Possible answer: The first two steps are the same.

2. What is different about solving the two kinds of equations?

   Equations with fractions have extra steps.

3. What is the first step to solve the problem $\frac{2}{5}w = 12$?

   Divide both sides by $\frac{2}{5}$.

4. What is the second step?

   Multiply 12 by $\frac{5}{2}$.

5. The value of *w* is $w = 30$

## Puzzles, Twisters & Teasers
### 5-10 It's All Just Words

Write these sentences as equations, solve the equations, and circle the answers. Put the letter of the correct answer on the spaces with the problem numbers. You will get 3 words that mean different things, but are all pronounced the same!

1. A number *n* is multiplied by one-third, giving a product of twelve.

   $n =$    4 B    36 **E**    15 L

2. Solve: A number *w* times one-third equals one-sixth.

   $w =$    $\frac{1}{3}$ P    2 J    $\frac{1}{2}$ **O**

3. Solve: The sum of *p* and three-sevenths gives one.

   $p =$    $\frac{4}{7}$ **U**    1 D    $\frac{1}{7}$ A

4. Solve: One-seventh times a number *t* equals one.

   $t =$    $\frac{1}{7}$ G    1 N    7 **F**

5. Subtracting three-fourths from a number *b* yields one and one-half.

   $b =$    $\frac{3}{4}$ S    $\frac{9}{4}$ **R**    $\frac{1}{2}$ H

This word is a number. $\underline{F}_4 \underline{O}_2 \underline{U}_3 \underline{R}_5$

This word is a preposition. $\underline{F}_4 \underline{O}_2 \underline{R}_5$

This word is used in the game of golf. $\underline{F}_4 \underline{O}_2 \underline{R}_5 \underline{E}_1$